Lecture Notes in Networks and Systems

Volume 568

The series "Lecture Notes in Networks and Systems" publishes the latest developments in Networks and Systems—quickly, informally and with high quality. Original research reported in proceedings and post-proceedings represents the core of LNNS.

Volumes published in LNNS embrace all aspects and subfields of, as well as new challenges in, Networks and Systems.

The series contains proceedings and edited volumes in systems and networks, spanning the areas of Cyber-Physical Systems, Autonomous Systems, Sensor Networks, Control Systems, Energy Systems, Automotive Systems, Biological Systems, Vehicular Networking and Connected Vehicles, Aerospace Systems, Automation, Manufacturing, Smart Grids, Nonlinear Systems, Power Systems, Robotics, Social Systems, Economic Systems and other. Of particular value to both the contributors and the readership are the short publication timeframe and the world-wide distribution and exposure which enable both a wide and rapid dissemination of research output.

The series covers the theory, applications, and perspectives on the state of the art and future developments relevant to systems and networks, decision making, control, complex processes and related areas, as embedded in the fields of interdisciplinary and applied sciences, engineering, computer science, physics, economics, social, and life sciences, as well as the paradigms and methodologies behind them.

Indexed by SCOPUS, INSPEC, WTI Frankfurt eG, zbMATH, SCImago.

All books published in the series are submitted for consideration in Web of Science.

For proposals from Asia please contact Aninda Bose (aninda.bose@springer.com).

Marvic Sciberras · Alexiei Dingli

Investigating AI Readiness in the Maltese Public Administration

Springer

Marvic Sciberras
IDEA Academy Malta
Mosta, Malta

Alexiei Dingli
University of Malta
Msida, Malta

ISSN 2367-3370 ISSN 2367-3389 (electronic)
Lecture Notes in Networks and Systems
ISBN 978-3-031-19899-1 ISBN 978-3-031-19900-4 (eBook)
https://doi.org/10.1007/978-3-031-19900-4

This Springer imprint is published by the registered company Springer Nature Switzerland AG
The registered company address is: Gewerbestrasse 11, 6330 Cham, Switzerland

Contents

Abbreviations

AI	Artificial Intelligence
AGI	Artificial General Intelligence
ANI	Artificial Narrow Intelligence
ASI	Artificial Superintelligence
CIOs	Chief Information Officers
DPO	Data Protection Office
EU27	Refer to the 27 European Countries
ICT	Information and Communications Technology
ICTs	Information and Communication Technologists
IMU	Information Management Unit
IT	Information Technology
MAFA	Ministry for Agriculture, Fisheries, Food and Animal Rights
MFED	Ministry for Education
MFH	Ministry for Health
MDIA	Malta Digital Innovation Authority
MITA	Malta Information Technology Agency
MTA	Malta Tourism Authority
OPM	Office of the Prime Minister
POYC	Pharmacy of Your Choice
WEF	World Economic Forum

List of Figures

List of Tables

Chapter 1
Introduction

The ubiquity of new technologies in our everyday lives has led to the lead-in of emerging technologies in the public sector. In a recent report by Microsoft and EY [1], entitled 'Artificial Intelligence in the Public Sector: European Vision for 2020 and Beyond', two-thirds of respondents rated AI as a digital priority. Although many local, state, and national governments see the promise of AI, only 4% of the organizations surveyed had adopted it for transformation. Consequently, out of all respondent organisations, 10% are using AI to solve complex problems, 9% are using it to substantially modify operating processes, while only 12% of the participants were using it to have a direct effect on the citizens and the broader economy.

In this regard, through the publication of '*Malta—The Ultimate AI Launchpad; A Strategy and Vision for Artificial Intelligence in Malta 2030*', the Maltese Government (2019) is committed to achieving ambitious goals in view of the oncoming AI revolution. The Government's proposition to entrepreneurs, along with foreign and local companies lies in the possibility to experiment, prototype, and ultimately test and scale AI discoveries through the Maltese Innovative Technology Certification and Legislation [2, 3]. To this extent, the public administration is also at the core of the Malta AI Strategy, whereby the AI endeavour is kicking off with six (6) high-level pilot projects across the different ministries [4–10]. Additionally, several proposals addressing ethical considerations, awareness campaigns and educational undertaking are also detailed out. An extraordinary investment in technology has been a major factor in the significant amount of work completed over the past nine years that has resulted in accessible public services wherever and at any time [11]. In view of the foregoing, and the changes the public administration is allegedly undergoing, it is apprehended that AI technologies will be redesigning the delivery of public services as well as work processes at the back-end of the national public service system. These are being reinforced with the launch of a new public service strategy, whereby 2022 to 2027 shall see a public service achieving a service of excellence. The 5-year strategy revolves around 3 societal pillars that are technology, service and people and

M. Sciberras and A. Dingli, *Investigating AI Readiness in the Maltese Public Administration*, Lecture Notes in Networks and Systems 568,
https://doi.org/10.1007/978-3-031-19900-4_1

3 public service principles namely quality, accountability and sustainability [12–14]. A bold future awaits the Maltese public administration embracing these values.

References

1. Microsoft (2020) Artificial intelligence in the public sector – European outlook for 2020 and beyond. https://info.microsoft.com/WE-DTGOV-CNTNT-FY21-09Sep-22-ArtificialIntelligenceinthePublicSector-SRGCM3835_01Registration-ForminBody.html. Accessed 21 May 2021
2. MDIA (2021a) Malta digital innovation authority. https://mdia.gov.mt/. Accessed 14 July 2021
3. MDIA (2021b) Malta's national AI strategy ranked 10th best amongst 54 countries. https://mdia.gov.mt/author/mdiaadmin/. Accessed 03 Jan 2021
4. Government of Malta (2019a) MALTA the utimate AI launchpad; A strategy and vision for artifical intelligence in Malta 2030. Government of Malta, Valletta
5. Government of Malta (2019b) Malta towards an AI strategy: high level policy document for public consultation. https://malta.ai/wp-content/uploads/2019/04/Draft_Policy_document_-_online_version.pdf. Accessed 12 June 2021
6. Government of Malta (2019c) Malta AI. https://malta.ai/. Accessed 10 Aug 2021
7. Government of Malta (2019d) Malta: the ultimate AI launchpad. https://malta.ai/wp-content/uploads/2019/11/Malta_The_Ultimate_AI_Launchpad_vFinal.pdf. Accessed 21 Aug 2021
8. Government of Malta (2019e) Malta: towards trustworthy AI: Malta's ethical AI framework. MDIA, s.l.
9. Government of Malta (2019f) Mapping tomorrow: a strategic plan for the digital transformation of the public administration 2019–2021. Office of the Principle Permanent Secretary (Office of the Prim Minister) and MITA, Valletta
10. Government of Malta (2019g) Mapping tomorrow; a strategic plan for the digital transformation fo the public admnistration 2019–2021. Government of Malta, Valletta
11. Government of Malta (2022) New public service strategy for the next 5 years is being implemented. https://publicservice.gov.mt/en/Pages/News/2022/20220503_PR220547.aspx. Accessed 01 Aug 2022
12. Government of Malta (2021a) New public service strategy. https://publicservice.gov.mt/en/pages/New_Public_Service_Strategy.aspx. Accessed 03 Oct 2021
13. Government of Malta (2021b) Press release by the offcie of the principal permanent secretary: Public service week 2021 kickstratrs today. https://www.gov.mt/en/Government/DOI/Press%20Releases/Pages/2021/May/28/pr211019en.aspx. Accessed 12 Aug 2021
14. Government of Malta (2021c) The public service and the public sector. https://publicservice.gov.mt/en/Pages/The%20Public%20Service/PublicServicePublicSector.aspx. Accessed 12 Sept 2021

Chapter 2
AI and Its Wider Impact

Intelligence research is one of the oldest fields of social sciences [1]. Philosophers have been attempting to explain how seeing, knowing, recalling, and reasoning works for over 2000 years [2]. The Greek philosopher Aristotle did not believe that technology could simply be used to mimic nature but argues that the major role of technology is to effectuate what nature cannot and in other cases to replicate nature's creation [3]. However, the European Parliament [4] discusses that despite the advantages that populations will benefit from Artificial Intelligence (AI), there are threats that need to be addressed to ensure accountability, safeguard human rights and minimise an adverse economic impact. Furthermore, in an article published by Edge Foundation, Poundstone [5] argues that in the realm of machines, almost anything that is conceptualised, physically conceivable, and fairly priced can be materialised. Thus, regardless of its practical use, human-like machine intelligence is a phenomenon with a proclaimed future. If technology is available, it will become increasingly affordable to enthusiasts, hackers, and "machine rights" movements and there will be interest in developing machines with unclear motives. This notion excludes the machinery that insurgents, rogue regimes, and intelligence services with crafty intentions may also develop. Hence, despite Aristotle's and fellow scientists' confidence in AI benefits domineering societal functions, the possibility that AI will rebel against its creators is one worth considering.

In this regard, no universally accepted definition of AI exists [6–8]. Artificial Intelligence is a diverse field of research that includes not just computer science but also psychology, economics, linguistics, and other disciplines, it aims to both create and comprehend intelligent entities [9]. The American Psychology Association defines 'intelligence' as the ability to gather knowledge, learn through experience and respond to changes. Moreover, comprehension of situations, the use of thought and applying the correct reasoning are also processes of 'intelligence' [10]. This school of thought could also be applied to AI.

M. Sciberras and A. Dingli, *Investigating AI Readiness in the Maltese Public Administration*, Lecture Notes in Networks and Systems 568,
https://doi.org/10.1007/978-3-031-19900-4_2

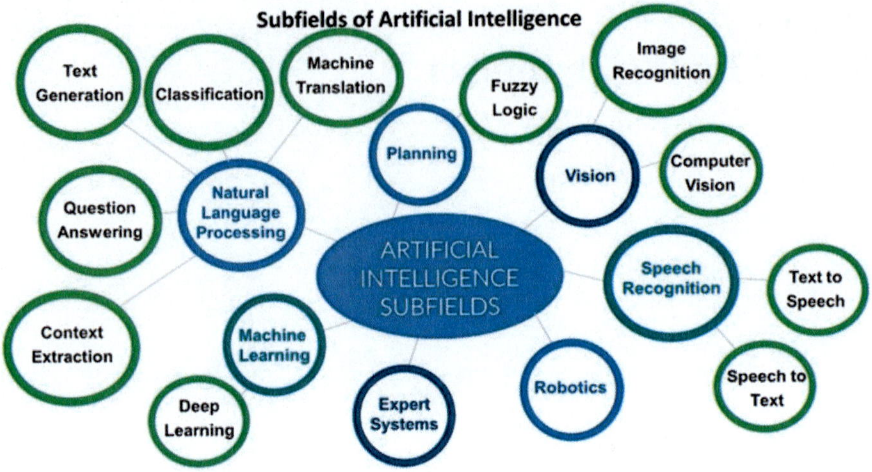

Fig. 2.1 Subfields of AI (*Source* Author's representation)

In 1950, Alan Turing, the founding father of the AI concept, illustrates 5 different methods that could be used to create a thinking machine, mainly through (i) programming (ii) machine learning (iii) logic and statistics (iv) cognitive learning and (v) context information [11]. John McCarty subsequently organised a workshop held in Dartmouth College in 1956 and the team eventually coined the term Artificial Intelligence. The participants discussed the emerging disciplines of computer science, natural language processing, and neural networks and AI was defined as the making of man-made intelligent machines through the application of science and engineering [12]. Fundamentally, AI can be categorised as analytical (cognitive), inspired by humans (emotional), and humanised (socially intelligent) according to the sort of intelligence it displays [13]. Furthermore, based on its evolutionary stage, AI can be classified as Artificial Narrow, General, and Super Intelligence [14].

Hence, AI can be distinguished in several subfields that constitutes of different techniques, applications and designs. These systems can work in isolation or be linked to other AI systems to deliver a specific outcome. Figure 2.1 illustrates several subfields of AI; however, it is to be noted that this not an exhaustive list.

To this extent, AI is reshaping human–machine interactions. When contemplating on how AI technology is invading our daily endeavours, one cannot disregard the notion that it effectively accentuates how humans are increasingly relying on the emerging assistive technology to enhance their lifestyles, monitor their wellbeing as well as for the provision of entertainment and communication [15]. Simply put, this disruptive technology is becoming a leader in our lives and humanity will need to face these new issues with nuanced methods, inquisitiveness, and caution in light of AI's expanding prevalence [16].

References

1. Carl N, Woodley of Menie MA (2019) A scientometric analysis of controversies in the field of intelligence research. Intelligence 77(101397):1–6
2. Norvig P, Russell SJ (1995) Artificial intelligence: a modern approach. Englewood Cliffs, Prentice Hall, New Jersey
3. Schummer J (2001) Artistotle on technology and nature. Philos Nat 38(1):105–120
4. European Parliment (2021) Artificial intellignece: threats and opportunities. https://www.europarl.europa.eu/news/en/headlines/society/20200918STO87404/artificial-intelligence-threats-and-opportunities. Accessed 02 May 2021
5. Poundstone W (2015) 2015: What do you think about machines that think?. https://www.edge.org/response-detail/26043. Accessed 02 May 2021
6. Ertel W (2018) Introduction to artificial intelligence, 2nd edn. Springer Nature, Cham
7. Monett D, Lewis CWP (2018) Getting clarity by defining artificial intelligence. In: Müller VC (ed) Philosophy and thepry of artificial intelligence. Springer, Berlin, pp 212–214
8. Nilsson NJ (2009) The quest for artificial intelligence: a history of ideas and achievements. Cambridge University Press, Cambridge
9. Wang P (2019) On defining artificial intelligence. J Artif Gen Intell 10(2):1–37
10. Khalfa J, Gullickson T (1995) What is intelligence? Contemp Ryschol APA Rev Books 40(7):708–709
11. Turing A (1950) Computing machinery and intelligence. Mind LIX(236):433–460
12. Gauglitz G (2019) Artificial vs. human intelligence in analytics. Anal Bioanal Chem 411:5631–5632
13. Haenlein M, Kaplan A (2019) A brief history of artificial intelligence: on the past, present and future of artificial intelligence. Calif Manage Rev 61(4):5–14
14. Brown S (2020) The innovation ultimatum: how six strategic technologies will reshape every business in the 2020s. Wiley, s.l.
15. Bhaumik A (2018) From AI to robotics: mobile, social and sentientrobots, 1st edn. CrC Press, Boca Raton, FL
16. Hoadley DS, Lucas NJ (2018) Artificial intelligence and national security. CRS Report, USA Congress

Chapter 3
Public Administration and Technology in Malta

The very first Information and Communications Technology (ICT) vision published by the Maltese Government dates back to 1994, of which document addresses policies for techno-economic growth and the components and processes necessary for a positive impact [1]. Interestingly, Camilleri argues how human capital, information technology (IT), active information exchange, grants to instigate growth in the private sector and sound policy structures were deemed to be the driving forces for an effective IT transition that registers fundamental economic growth for the Maltese islands.

Essentially, IT underpins the whole spectrum of computing and communication technology used by an organisation, both internal and external, to make critical decisions and improve productivity against targets. These computer and communication systems are made up of a variety of hardware and software components that are used to handle data [2]. In this regard, data-collecting devices, data processing equipment, output devices, storage system, communications equipment, and software are all included in an information management system architecture [3].

Consequently, the publishing of this strategy solidified the Malta Information Technology Agency (MITA) which institution was established by the government in 1990 to administer IT processes government-wide. To date, besides providing the Government with ICT infrastructures, systems and services utilising contemporary technology platforms, but also integrating technological advancements in concrete and tangible business solutions assuring the Government's plans and projects are skilfully implemented and return on investment is maximised [4].

Fast forward twenty (20) years, the Government has now established Information Management Units (IMU) within each ministry, which engages the required personnel to address ICT matters and logistics. In 2014, a new National Digital Strategy is published, envisaging a robust, innovative and world-class digital economy [5, 6]. *Digital Malta* addressed three (3) strategic subjects which include the digital citizen, digital business and digital government that are structured through a legislative framework, sound infrastructure and enhanced human capital. To this

M. Sciberras and A. Dingli, *Investigating AI Readiness in the Maltese Public Administration*, Lecture Notes in Networks and Systems 568,
https://doi.org/10.1007/978-3-031-19900-4_3

effect, the execution of this digital vision brought about socio-economic benefits through the provision of sustainable services while creating the ideal economic environment to stimulate investment in technology and human resources. To this effect, digital transformation and diversity of services remain significant for generating a sturdy national economy [7]. Additionally, digitising the public administration has been a major milestone within the government during recent years.

Mapping Tomorrow [8] was a strategy designed to address further digital transformations within the Maltese public administration between 2019 and 2021. It established the Government's objective to provide high-quality public services that are constructed around convenience and anytime accessibility. Effectually, technology has taken on a transformational role in the design and delivery of Maltese public services. Notably, during the last decade, Malta has ranked high in the e-Government Benchmark issued yearly by the European Commission. The 2020 report ranked Malta as the lead country, scoring 97% on the overall country performance [9–18]. Addressing the public administration needs with foresight, the strategy acknowledges that data exchange and trust-building are two cornerstones for citizen-centric services. Additionally, through a connected government and the elimination of silos, the 'Once-Only Principle' can be achieved, of which initiative was launched by the European Commission in 2017 as part of the eGovernment Action Plan 2016–2020 [19]. The vision of *Mapping Tomorrow* was to create a continually renewed digital public administration that transforms how individuals and corporations interact with the Government while delivering exceptional public services.

The continuum of *Mapping Tomorrow* is being ingrained with a new public service strategy launched by the Principal Permanent Secretary, Mr Mario Cutajar, in November 2021 [20]. The strategy will be set for five (5) years and its three (3) focal elements are the people, technology and service. Three (3) different working groups have been assigned to thoroughly address each component of the strategy ensuring a detailed yet holistic approach to the strategy. The driving force of the strategy is set on quality deliverance, accountability and sustainability.

References

1. Camilleri J (1994) A national strategy for information technology for Malta, 1st edn. University of Malta, Msida
2. Sinha PK, Sinha P (2016) Information technology: theory and practice, 1st edn. PHI Learning Private Ltd., Haryana
3. Ulrich WM, Newcomb P (2010) Information systems transformation: architecture-driven modernization case studies, 1st edn. Morgan Kaufmann Publishers, Burlington, MA
4. MITA (2021) MITA. https://mita.gov.mt/about-us/. Accessed 09 June 2021
5. Government of Malta (2014a) Digital Malta 2014–2020. https://digitalmalta.org.mt/en/Documents/Digital%20Malta%202014%20-%202020.pdf
6. Government of Malta (2014b) Digital Malta: national digital strategy 2014–2020. Government of Malta, s.l.
7. Guarise G (2021) Malta is ready to restart from the digital economy. https://www.maltabusiness.it/en/malta-is-ready-to-restart-from-the-digital-economy/. Accessed 16 June 2021

8. PPS (2019) Mapping tomorrow. Office of the Principle Permanent Secretary, Valletta
9. EU Commission (2021a) A European strategy for data. https://digital-strategy.ec.europa.eu/en/policies/strategy-data. Accessed 06 May 2021
10. EU Commission (2021b) Coordinated plan on artificial intelligence 2021 review. https://digital-strategy.ec.europa.eu/en/library/coordinated-plan-artificial-intelligence-2021-review. Accessed 12 Sept 2021
11. Commission EU (2021) Digital education action plan 2021–2027. EU Commission, Brussels
12. EU Commission (2021d) eGovernment Benchmark 2020: eGovernment that works for the people. https://digital-strategy.ec.europa.eu/en/library/egovernment-benchmark-2020-egovernment-works-people. Accessed 12 June 2021
13. EU Commission (2021e) Excellence and trust in AI – brochure. https://digital-strategy.ec.europa.eu/en/library/excellence-and-trust-ai-brochure. Accessed 15 June 2021
14. EU Commission (2021f). Fostering a European approach to artificial intelligence. COM(2021) 205 final, Brussels
15. EU Commission (2021g) Shaping Europe's digital future. https://digital-strategy.ec.europa.eu/en/news/new-report-looks-ai-national-strategies-progress-and-future-steps. Accessed 10 June 2021
16. EU Commission (2021h) Strategy for artificial intelligence. https://digital-strategy.ec.europa.eu/en/policies/strategy-artificial-intelligence. Accessed 06 May 2021
17. EU Commission (2021i) The digital Europe programme. https://digital-strategy.ec.europa.eu/en/activities/digital-programme. Accessed 15 06 2021
18. EU Commission (2021j) The digital Europe programme. https://digital-strategy.ec.europa.eu/en/activities/digital-programme. Accessed 06 May 2021
19. Krimmer R (2021) The once only principle. https://toop.eu/. Accessed 12 June 2021
20. Public Service Malta (2021) New strategy for the public service – achieving a service of excellence. https://publicservice.gov.mt/en/Pages/Initiatives/Achieving-a-Service-of-Excellence.aspx. Accessed 11 Apr 2022

Chapter 4
AI Adoption

The Institute for Competitiveness published a report which investigates why the EU is currently at a competitive disadvantage when compared to the US and China when it comes to AI technology and investment [1]. However, the European Commission is guiding the EU27 in how best to implement AI with "an ethical and trustworthy" approach [2]; this will provide the Member States with a maximised potential to better exploit AI while tackling challenges collectively [3]. To complement this position, a Coordinated Plan on AI puts forth specific ideas and recommendations for Member States cooperative action to enhance European competitivity in the global AI environment [4–13]. Through the data economy and its applications as defined in the European Data Strategy [4–13], Europe aims to merge science and technology strengths with high-quality digital infrastructure and a legislative framework. The vision to become a worldwide leader is based on its key values, that will therefore establish a sound AI ecosystem, bringing technological benefits to the society and economy of Europe as a whole [4–13].

In order to actualise this all-encompassing European AI-vision [14–16], Member States compiled national strategies in line with the EU Commission's guidelines and prioritised AI investment in their political agendas. To target these aspects, considerable efforts are devoted to promoting an inclusive digital transformation, based on core European values, such as transparency and non-discrimination, while setting an economically competitive environment. The European Union is committed to investing €20 billion till 2030 [4–13], to incentivise both the private organisations and Member States to capitalise in AI initiatives, to be at the forefront of competition and to harness talent Europe-wide. Through the research and innovation programme Horizon 2020, the commission is envisaging a strengthened connection to AI research across the Member States through the development of AI platforms and system applications tailored to key policy areas.

In this regard, the Government of Malta seized the opportunity of such direction to draw an ethical framework specifying how AI adoption will fall within regulatory processes. The central objectives of *Malta Towards a Trustworthy and Ethical AI*

M. Sciberras and A. Dingli, *Investigating AI Readiness in the Maltese Public Administration*, Lecture Notes in Networks and Systems 568,
https://doi.org/10.1007/978-3-031-19900-4_4

are to identify need for AI to operate in a human-centric environment, exercise respect to the laws of Malta, reap AI benefits whilst controlling their risks and practice AI adoption in line with globally set ethical standards [17–23]. Furthermore, a national AI strategy was published to upsurge AI adoption nationwide. The strategy, identified as *Malta: The Ultimate AI Launchpad—A strategy and vision for Artificial Intelligence in Malta 2030* [17–23] (Malta National Strategy) ranked 10th on the Global AI Index amongst 54 countries [24, 25], surpassing other EU countries whose AI adoption is better established.

Leading from this end, the strategy establishes the vision of the Government with the aim for Malta to create an environment whereby AI systems can be tested and henceforth implemented worldwide; essentially, to become the world-leading facilitator of AI technologies. The strategy was devised through the Malta. AI Taskforce whose differentiated backgrounds, including law representing individuals, businessmen and academics, public governing bodies and technology experts, proved fundamental in formulating an overarching all-inclusive strategy. The task force identified the strategic pillars and enabling factors of which one without the other will hinder AI adoption. Educating the workforce and the Maltese population about AI and its uses, ensuring an ethical regulatory framework is in practice to safeguard privacy and security, and a robust infrastructure to cater for the volumes of data generation are the main factors that will enable a smooth adoption and transition into an advanced AI era. Moreover, the Government is committed to research how AI will affect the Maltese labour market in order to provide the workforce with the new digital skills needed to remain competitive. Thereupon, the strategy targets the public and private sectors through the investment in innovative start-ups and research and design (R&D) entities. Facilitating an ideal environment through proper support structure and available co-funding opportunities are set to attract talent, newly established organisations and scale-ups to invest in AI systems and R&D activities in Malta [17–23]. Furthermore, as indicated by the EU Commission, in order to address the change, the AI revolution will bring forth, the restructuring of education systems is necessary. The Malta AI Strategy details the Government's plan on this matter and identifies which educational stages could rapidly take advantage from such adoption. As such, it is expected that educators, administrators and students will reap immediate benefits from the adoption of an adaptive learning systems. Through the customisation of personalised learning, it is aimed to positively influence the teaching and learning journeys. In addition, the provision of a business intelligence is proposed to identify potential early school leavers and address their needs effectively. Throughout all this, students and parents are the focus of the process as well as educators who will be the driving force behind these educational-AI initiatives.

The EU Commission is also targeting the re-training and upskilling of the current public administrators, to prepare the workforce with the technical skills needed, as outlined in the Digital Education Action Plan [4–13]. In line with this, the Malta AI Strategy outlines processes that are going to be undertaken to upskill the Maltese public workforce. These include the launch of an awareness campaign, that would better inform the public administrators of AI at their workplace as well as a selection of AI and ICT related courses to foster knowledge and understanding of AI processes.

To this effect, the Malta AI Strategy targets five further key policy areas whereby AI systems will be piloted and studied with the target that eventually they will be adopted nationwide. The transport project includes the adoption of a traffic management system through the use of a machine learning system to control traffic flow in identified traffic junctions, updating the geographic information system of the Maltese islands to have real-time visibility of public transport and undertake a research study to identify possible nation-wide enhancements as part of the traffic management system. Considerably, the deployment of AI systems on the roads has the potential to increase pedestrian safety, assist in traffic management and monitor passenger-flow management to provide constructive feedback for decision-making purposes [26].

One of the public policy areas that are investing heavily in emerging technologies is the healthcare sector [27]. The fast development in AI has allowed aggregate healthcare data to build sophisticated models that can automate diagnoses [28] and enable a more precise approach to medicine through the timely and dynamic adaptation of treatments with the highest effectiveness [29]. Additionally, machine learning algorithms can facilitate appointment scheduling by prioritising according to the patients' conditions and needs, minimising delays with increased efficiency [30]. Medical imaging is also a highly ranked AI activity due to the widespread usage of standard image formats, which provide appropriate databases to train AI systems on. In this field, the pilot project identified by the Government as fundamental for an enhanced public service, is the Pharmacy of your Choice platform, whereby the qualified personnel have real-time information at hand to make informative decisions and assist the individual seeking assistance. Additionally, the system will identify possible savings and help draw up diverse healthcare approaches through anticipating societal needs. As a contact-centre representative, mechanically intelligent AI gives predetermined solutions to trivial client issues, whereas analytically intelligent AI examines customer difficulties [31]. Applying this concept, the Maltese government is commissioning another citizen-centric initiative, managed by the servizz.gov team. The initiative involves the adoption of AI-powered email assistants to enhance the workflow of the servizz.gov team while providing a 24/7 public service. Fast forward, the government foresees chat-bot interaction that will assist the citizens in gaining the information they need and the provision of translation to facilitate communication with foreign nationals [17–23].

The advancement of technology has established a new category of tourism known as Tourism 4.0 [32], whose primary features are swiftness and prevalence. Tourism 4.0 links information available online with smart technologies to enhance the visitor's travel plans and activities [33]. In Malta, tourism is one of the pillars contributing to the country's economy; in fact, the Eurostat reports Malta as the state with the highest tourism intensity in Europe [34] and 2019 as the lead in hotel occupancy at 75% [35, 36]. To anticipate and meet the demands of tourists, the government will set up an AI-powered platform that would present the user with activities specific to their interests. The platform will be based on social media principles, whereby according to the user's searches, the platform will match events and display similar recreational suggestions. In addition, through the user's movements across the island, the AI model will be able to predict future planned holiday activities and be able to present real-time recommendations.

The introduction of green energy as part of the EU's agenda, and the need to look into more renewable and sustainable energy resources [14–16], calls for the Maltese government to ensure that simplified but effective processes are in place concerning data collection and analysis approach. To facilitate this, the country's utility departments will join forces in implementing AI systems that through data-mining, AI can assist in providing just-in-time possible outcomes. Furthermore, the systems will be developed to help identifying issues and pre-empting actions to minimise time-consuming procedures [17–23].

As a digital technology that has the ability to profoundly alter the public sector, AI stimulates the public administration's dual potential for economic and social growth, which is to become the newest trend in public services development and fosters governments reform [37, 38]. To this extent, the Maltese government is committed to facilitating for the necessary change so that Malta can be recognised as an AI-driven society [17–23].

References

1. Compagnucci S et al (2020) The way to digital made-in-Europe. Promoting European values in the global digital arena. https://www.i-com.it/en/2020/12/08/the-way-to-digital-made-in-europe-promoting-european-values-in-the-global-digital-arena-2/. Accessed 04 May 2021
2. AI HLEG (2019) Ethics guidelines for trustworthy AI. https://digital-strategy.ec.europa.eu/en/library/ethics-guidelines-trustworthy-ai. Accessed 15 June 2021
3. Samoilo S et al (2020) AI watch: defining artificial intelligence. Publications Office of the European Union, Luxemburg
4. EU Commission (2021a) A European strategy for data. https://digital-strategy.ec.europa.eu/en/policies/strategy-data. Accessed 06 May 2021
5. EU Commission (2021b) Coordinated plan on artificial intelligence 2021 review. https://digital-strategy.ec.europa.eu/en/library/coordinated-plan-artificial-intelligence-2021-review. Accessed 12 Sept 2021
6. EU Commission (2021c) Digital education action plan 2021–2027. EU Commission, Brussels
7. EU Commission (2021d) eGovernment benchmark 2020: eGovernment that works for the people. https://digital-strategy.ec.europa.eu/en/library/egovernment-benchmark-2020-egovernment-works-people. Accessed 12 June 2021
8. EU Commission (2021e) Excellence and trust in AI – brochure. https://digital-strategy.ec.europa.eu/en/library/excellence-and-trust-ai-brochure. Accessed 15 June 2021
9. EU Commission (2021f) Fostering a European approach to artificial intelligence. COM(2021) 205 final, Brussels
10. EU Commission (2021g) Shaping Europe's digital future. https://digital-strategy.ec.europa.eu/en/news/new-report-looks-ai-national-strategies-progress-and-future-steps. Accessed 10 July 2021
11. EU Commission (2021h) Strategy for artificial intelligence. https://digital-strategy.ec.europa.eu/en/policies/strategy-artificial-intelligence. Accessed 06 May 2021
12. EU Commission (2021i) The digital Europe programme. https://digital-strategy.ec.europa.eu/en/activities/digital-programme. Accessed 15 June 2021
13. EU Commission (2021j) The digital Europe programme. https://digital-strategy.ec.europa.eu/en/activities/digital-programme. Accessed 06 May 2021

14. European Commission (2021a) A European approach to artificial intelligence. https://digital-strategy.ec.europa.eu/en/policies/european-approach-artificial-intelligence. Accessed 17 June 2021
15. European Commission (2021b) Ethics guidelines for trustworthy AI. https://ec.europa.eu/futurium/en/ai-alliance-consultation.1.html. Accessed 19 June 2021
16. European Commission (2021c) Sustainable development goals. https://ec.europa.eu/info/strategy/international-strategies/sustainable-development-goals_en. Accessed 03 July 2021
17. Government of Malta (2019a) MALTA the utimate AI launchpad; A strategy and vision for artifical intelligence in Malta 2030. Government of Malta, Valletta
18. Government of Malta (2019b) Malta towards an AI strategy: high level policy document for public consultation. https://malta.ai/wp-content/uploads/2019/04/Draft_Policy_document_-_online_version.pdf. Accessed 12 June 2021
19. Government of Malta (2019c) Malta AI. https://malta.ai/. Accessed 10 Aug 2021
20. Government of Malta (2019d) Malta: the ultimate AI launchpad. https://malta.ai/wp-content/uploads/2019/11/Malta_The_Ultimate_AI_Launchpad_vFinal.pdf. Accessed 21 Aug 2021
21. Government of Malta (2019e) Malta:towards trustworthy AI: Malta's ethical AI framework. MDIA, s.l.
22. Government of Malta (2019f) Mapping tomorrow: a strategic plan for the digital transformation of the public administration 2019–2021. Office of the Principle Permanent Secretary (Office of the Prim Minister) and MITA, Valletta
23. Government of Malta (2019g) Mapping tomorrow; A strategic plan for the digital transformation fo the public admnistration 2019–2021. Government of Malta, Valletta
24. MDIA (2021a) Malta digital innovation authority. https://mdia.gov.mt/. Accessed 14 July 2021
25. MDIA (2021b) Malta's national AI strategy ranked 10th best amongst 54 countries. https://mdia.gov.mt/author/mdiaadmin/. Accessed 03 July 2021
26. Joshi N (2019) How AI can transform the transportation industry. https://www.forbes.com/sites/cognitiveworld/2019/07/26/how-ai-can-transform-the-transportation-industry/?sh=26d02f264964. Accessed 15 June 2021
27. Yang Z, Ng B-Y, Kankanhalli A, Luen-Yip JW (2012) Workarounds in the use of IS in healthcare: a case study of an electronic meication administration system. Int J Hum Comput Stud 70:43–65
28. Keane PA, Topol EJ (2018) With an eye on AI and autonomous diagnosis. NJP Digit Med 1(40)
29. Saria S (2014) A $3 trillion challenge to computational scientists: transforming healthcare delivery. IEEE Intell Syst 29:82–87
30. Nelson A, Herron D, Rees G, Nachev P (2019) Predicting scheduled hospital attendance with artificial intelligence. Digit Med 2(1):1–7
31. Rust RT, Huang MH (2021) The feeling economy: how artificial intelligecne is creating the era of empathy, 1st edn. Palgrave Macmillan, Cham
32. Pencarelli T (2019) The digital revolution in the travel and tourism industry. Inf Technol Tour 22(3):455–476
33. Kourouthanassis PE, Mkalef P, Pappas IO (2017) Explaining travellers online information satisfaction: a complexity theory approach on information needs, barriers. sources and personal characteristics. Inf Manage 54(6):814–824
34. EU Commission (2017). Euroststcelebrates Malta. https://ec.europa.eu/eurostat/web/products-eurostat-news/-/edn-20170921-1. Accessed 16 June 2021
35. Eurostat (2020a) European construction sector observatory: country profile Malta. https://ec.europa.eu/growth/sites/default/files/ecso_cfs_mt_2021.pdf. Accessed 16 June 2021
36. Eurostat (2020b) Tourism statistics – annual results for the accommodation sector. https://ec.europa.eu/eurostat/statistics-explained/index.php?title=Tourism_statistics_-_annual_results_for_the_accommodation_sector&oldid=541904#Continuous_growth_in_the_tourist_accommodation_sector. Accessed 09 June 2021

37. Valle-Cruz D, Ruvalcaba-Gomez AE, Sandoval-Almazan R, Criado I (2019a) A review od artificial intelligence in government and its potential from a public policy perspective. In: Proceedings of the 20th annual international conference on digital government reserach, pp 91–99
38. Valle-Cruz D, Ruvalcaba-Gomez AE, Sandoval-Almazan R, Criado IJ (2019b) A review of artificial intelligence in government and its potential from a public policy perspective. Association for Computing Machinery, Dubai, UAE

Chapter 5
AI-Readiness

As AI adoption in the Maltese islands is in its infancy, there is minimal literature to which one can refer to when researching the implementation of AI solutions in the Maltese public administration. In this regard, a report published by Capgemini Consulting, analysing the AI readiness in numerous countries of which the majority are European, ranked Malta the 19th out of 35 countries [1]. According to the benchmark report, Malta's readiness performs best on the IT side of operations and needs to instate more commitment on getting the private and public institutions prepared in terms of applying policy and the required support. Furthermore, Malta is lacking the advanced skills needed to design and analyse AI systems; this is fundamental to have specialists focusing on the advancement of AI technologies in the country.

Oxford University [2] defines readiness as a state of preparedness for something or the attitude to be willing to do something. In this regard, digital readiness is described by Nguyen et al. [3] as organisational preparedness in the face of digital transformation, in terms of technical and financial resources, managerial support, company culture, commitment, goal communication, and willingness to collaborate [4]. Hence, digital readiness can be categorised as organisational assets, skills and capabilities and commitment to change [5].

As discussed by Hanefi [6], AI-enabled activities significantly contribute to digital transformations within organisations, as these multifunctional emerging technologies execute researching functions, reasoning, problem-solving and decision making as well as apply perception through learning, analytical thinking and prediction through precise performance. In effect, AI is already manifesting potential applications that are fundamentally altering the convectional work-related ideologies and governments have already begun to incorporate AI innovations into their operations and service delivery to increase efficiency, save time and money, and provide higher-quality public services [7]. In the attempt to measure AI readiness, the *2019 Government AI Readiness Index* report assesses readiness in terms of data and infrastructure, education and skills, governance as well as government and public services. These are based on eleven (11) metrics to capture a detailed picture of the government AI readiness

M. Sciberras and A. Dingli, *Investigating AI Readiness in the Maltese Public Administration*, Lecture Notes in Networks and Systems 568,
https://doi.org/10.1007/978-3-031-19900-4_5

status quo. Following the 2019 index, the *Government AI Readiness Index 2020*, also published by Oxford Insights [8] investigates AI readiness based on thirty-three (33) indicators addressing the government and ethical governance, technology, human capital and innovation adaptability, and data infrastructure amongst others. Hence, it can be surmised readiness is evaluated in terms of the organisation's preparedness to withstand innovation and the institution's readiness to facilitate innovation [4].

As observed through the various levels of enquiring applied for the Oxford Insights index reports, AI readiness is a theoretical concept that differs in its connotation according to the contextual and conceptual application [9]. In addition, due to AI's multi-functional applications, it is sometimes unclear for organisations to identify the technology that will assist in achieving their goal [10]; as a result, implementing AI presents organisational, technological, and individual obstacles [11]. The capacity to redesign management systems and organisational structures, in response to new possibilities and challenges, will be critical for institutions to adapt to changing environments [12]. Consequently, while the majority of companies realise that AI is a transformative technology with enormous potential, a cautious approach is being adopted.

Consequently, the adoption of AI is a substantial interference in the administration of an organisation. It impacts a wide variety of fluctuating factors such as an organisation's culture, policy and procedures, risk management, strategic planning and views about the benefits of innovative applications [13]. Relatively, the readiness of an organisation to change may be described as the state of being ready and able to act [14]; however, digital transformation is the strategic transformation of all elements of the organisation underpinned by the establishment of a new ecosystem in which technology produces and generates value to stakeholders while facilitating the firm with flexibility and information to react more rapidly to changing conditions [15]. However, a critical consideration in developing and implementing an AI strategy is that plans may soon become obsolete as innovative disruptions continually emerge. Moreover, adopting new technology may require an organisation to alter its internal procedures, which is frequently challenging and may not always yield the desired results [16]. Designing and developing AI solutions is frequently an IT resource-intensive endeavour and integrating AI artefacts into an institution's existing IT infrastructure and landscape adds another layer of complexity [17].

As such, infrastructural preparedness necessitates the establishment, definition, and design of processes that enable the control and governance of AI systems to be successfully deployed [18]. Furthermore, [19] discuss that while governmental institutions adopt point solutions, they may also revamp their data and technology infrastructures to lay a better foundation for future AI applications. Moreover, it is argued that when organisations adopt AI for specific use cases, it is frequently discovered that these initiatives demonstrate scalable advantages, which may foster strong advocacy for AI and its wider potential. The success or failure rates established at this level ascertains the tone for future deployments of AI. Likewise, it is commonly accepted that AI is only as effective as the data on which it is developed, and it has an insatiable appetite for data. To administer data effectively, the governments must design a data governance system that incorporates engineering and reliable security while addressing requirements for data sourcing, access, and quality control [20]. Shearer et al. [8] reaffirm emphasising that AI tools have to be designed, developed and trained on high quality and representative data. They also highlight the importance of installing the proper infrastructure for effective use and service delivery.

On the other hand, technical readiness looks into the human capital aspect that would drive innovation. In this regard, leaders play an important role in promoting innovation and instilling a positive attitude towards emerging trends that would facilitate the workflow of the team [21]. Additionally, human capital refers to the abilities, expertise, and credentials of an individual, group, or workforce that are viewed as economic assets [22]. McNeill [23] defines skills as the specific learned abilities needed to execute a job successfully whereas competencies are the person's expertise, etiquette and attitude that lead them to a prosperous career. According to the WEF's *The Future of Jobs Report 2018* [24] references the adoption of new technology and advancement of AI technology-related skills as key to positive growth, while also outlining the need to develop soft skills such as creativity, critical thinking and persuasion as another part of the acquisition of the required skills. Hence, understanding change obstacles may help executives not only communicate with their workforce more effectively, but also identify which AI efforts are most viable, what training should be given, and what incentives may be necessary [25].

The transitioning to AI adoption by the government calls for the development of personal plans that can be established with the goal of upskilling or retraining employees in response to the shift in skills and competencies required as a result of AI's effect [26]. Moreover, to design, develop and implement robust AI solutions, it is necessary to nurture and recruit highly competent AI personnel while offering training and grant programs to the workforce who ultimately are seen as a critical segment for AI adoption [8]. In this regard, the Maltese government is committed to conducting research to determine which talents and occupations in society are more susceptible to the AI revolution as well as launching several incentives targeting the public administrators to re-train and upskill as required.

Furthermore, the preparedness of an organisation's workforce to embrace new technology in response to external influences is referred to as environmental readiness [27]. In this regard, Bateson [28] discusses that any adaptation process begins with an individual's approach and desire (or lack thereof) to modify an element of themselves in light of the contexts in which they operate. A growth mindset is a fundamental component to ensure adaptive abilities because of change. In addition, Bateson suggests that procedures for adapting AI must incorporate compassion and authenticity, embodiment, basic needs and aspirations and learn as well as consideration of what lay outside the organisation's borders. To this extent, change management processes enable people to comprehend and adapt to organisational change brought about by AI. Change management, in particular, is critical for dispelling misunderstandings about AI and the uncertainty related to job insecurity and layoffs [25]. This is critical given that often, AI solutions do not substitute roles in their entirety, but it tends to automate mundane activities or simplify process stages [29]. Employees and consumers must be educated about the benefits of AI to enhance acceptability.

Furthermore, AI adoption will call for increased collaborative environments whereby domain experts, AI professionals, and IT divisions interact and cooperate in cross-functional teams. Collaborative work is critical to overcome segregated processes and to find new systems that can be applied at the benefit the company. Thus, by encouraging different collaboration scenarios, organisations will have different skill sets complementing one another [30].

References

1. Tinholt D (2018) Artificial intelligence readiness and performance benchmark. https://www.capgemini.com/2018/07/artificial-intelligence-readiness-and-performance-benchmark/. Accessed 08 June 2021
2. Oxford University (2021) Oxford learner's dictionaries. https://www.oxfordlearnersdictionaries.com/definition/english/readiness. Accessed 13 June 2021
3. Nguyen DK, Broekhuizen T, Dong JQ, Verhoef PC (2019) Digital readiness: construct development and empitical validation. In: International conference in information systems, issue association for information systems (AIS), p 2966
4. Lokuge S, Sedera D, Grover V, Dongming X (2018) Organizational readiness for digital innovation: development and empirical calibration of a construct. Inf Manage 56:445–461
5. Pare G, Sicotte C, Poba-Nzaou P, Balouzakis G (2011) Clinicians' perception of organizational readiness for change in the context of clinical information system projects: insights from two cross sectional surveys. Implement Sci 6(1):15
6. Hanefi M (2020) The role of artificial intelligence within the scope of digitial tranfromation in enterprises. In: Ekren G, Erkollar A, Oberer B (eds) Advanced MIS and digital transformation for increased creativity anf innovation in business. IGI Global, Hershey PA, pp 122–146
7. Miller H, Stirling R (2019) Government artificial intelligence readines index 2019. Oxford Insight, Canada
8. Shearer E, Stirling R, Pasquarelli W (2020) Government AI readiness index 2020. Oxford Insights, Canada
9. Alsheibani S, Cheung Y, Messom C (2018) Artificial imtelligence adoption: AI-readiness ar firm-level. https://aisel.aisnet.org/pacis2018/37. Accessed 14 June 2021
10. Iansiti M, Lakhani KR (2020) Competing in the age of AI. https://hbr.org/2020/01/competing-in-the-age-of-ai. Accessed 14 June 2021
11. Bughin J et al (2017) Artificial intelligence: the next digital frontier. https://www.mckinsey.com/~/media/mckinsey/industries/advanced%20electronics/our%20insights/how%20artificial%20intelligence%20can%20deliver%20real%20value%20to%20companies/mgi-artificial-intelligence-discussion-paper.ashx. Accessed 14 June 2021
12. Chernov A, Chernova V (2019) Artificial intelligence in management: challenges and opportunities. In: 38th internatioanl scientific conference on economic and social development, Rabat
13. Inandi Y, Gilic F (2016) Relationship of teachers' readiness for change with their participation in decision making and school culture. Educ Res Rev 11(8):823–833
14. Oostemdorp LJ, Durand MA, Lloyd A, Elwyn G (2015) Measuring organisational readiness for patient engagement (MORE); an international online Delphi consensus study. BMC Health Serv Res 15(1):1–13
15. Ross J (2019) Digitial success requires breaking the rules. https://sloanreview.mit.edu/article/digital-success-requires-breaking-rules/. Accessed 16 June 2021
16. Gans J (2016) The disruption dilemma. MIT Press, Cambridge, MA
17. Hechler E, Oberhofer M, Schaeck T (2020) The operationalization of AI. In: Berkerley CA (ed) Deploying AI in the enterprise. Apress, pp 115–140

18. Eggers WD, Schatsky D, Viechnicki P (2017) AI-augmented government: using cognitive technologies to redesign public sector work. Deloitte University Press, s.l.
19. van Buren E, Chew B, Eggers WD (2020) AI readiness for government. Are you ready for AI?. https://www2.deloitte.com/us/en/insights/industry/public-sector/ai-readiness-in-government.html. Accessed 17 June 2021
20. Sohail O, Sharma P, Cric B (2018). Data governance for next generation platforms. https://www2.deloitte.com/content/dam/Deloitte/us/Documents/technology/us-big-data-governance.pdf. Accessed 18 June 2021
21. Heavin C, Power DJ (2018) Challenges fir digital transformation – towards a conceptual decision support guide for managers. J Decis Syst 27(1):38–45
22. Merriam-Webster (2021) Definition of human capital. https://www.merriam-webster.com/dictionary/human%20capital. Accessed 18 June 2021
23. McNeill J (2019) Skills vs competencies – What's the difference, and why should you care. https://social.hays.com/2019/10/04/skills-competencies-whats-the-difference/. Accessed 18 June 2021
24. WEF (2018) The future of jobs report 2018. https://www.weforum.org/reports/the-future-of-jobs-report-2018. Accessed 18 June 2021
25. Fountaine T, McCarthy B, Saleh T (2019) Building AI-powered organizations. Harvard Bus Rev July–August:1–13
26. Hoque E (2020) Upskilling the future workforce using AI and affective computing. In: Companion publication of the 2020 international conference on multimodal interactions, Netherlands, p 456
27. Yang ZJ, Sun J, Zhang YL, Wang Y (2015) Understanding SaaS adoption from the perspective of organizational users: a tripod readiness model. Comput Hum Behav 45:254–264
28. Bateson N (2016) Small arcs of larger circles framing through other patterns, 1st edn. Triarchy Press, Axminster
29. McAfee B, McAfee A (2017) The business of artificial intelligence. https://hbr.org/2017/07/the-business-of-artificial-intelligence. Accessed 18 June 2021
30. Davenport T (2018) From analytics to artificial intelligence. J Bus Anal 1:73–80

Chapter 6
Research Purpose, Question and Design—Exploratory Sequential Approach

The research purpose resulted from the launch of *Malta: The Ultimate AI Launchpad—A National AI Strategy* [1–7] and thereby, the goal of this study is to look at how equipped the Maltese government is for AI adoption in terms of transformational, organisational, technical and environmental readiness. Consequently, the research question guiding this study is:

> What is the extent of AI readiness of the Maltese Public Administration?

Furthermore, the following sub-question has been identified:

> What are the changes necessary for the successful adoption of AI technologies in the Maltese Public Administration?

Following this, the study began with a review of the literature to acquire deeper understanding of how AI is being employed by foreign public administration and to understand the benefits and constraints that AI adoption presents to the end-users. Following that, a qualitative study was undertaken to acquire insight and establish perceptions from the main AI facilitators within the Maltese public service. Subsequently, a quantitative survey was disseminated to gather the views of the public servants on the matter. Both outcomes were individually assessed and later cross-examined, giving depth to the research and its results.

Due to such research structure, and because the research question, goals, and objectives of this study are exploratory, the investigation employs an exploratory sequential approach. The sequential design divides the process into two stages by initially collecting qualitative data, followed by the compilation of information through a quantitative approach [8]. The methodology is beneficial when the main source of information is qualitative data and this design is particularly well-suited to investigating a phenomenon for which no guiding framework or theory exists and no prior assessments are available [9], which is the case with investigating AI readiness in the Maltese public administration.

M. Sciberras and A. Dingli, *Investigating AI Readiness in the Maltese Public Administration*, Lecture Notes in Networks and Systems 568,
https://doi.org/10.1007/978-3-031-19900-4_6

Finally, through the review of literature and exploration of the key concepts and their relevance in the context of the research question, a set of recommendations on the best practices to adopt AI within the Maltese public administration will be put forth.

References

1. Government of Malta (2019a) MALTA the utimate AI launchpad; A strategy and vision for artifical intelligence in Malta 2030. Government of Malta, Valletta
2. Government of Malta (2019b) Malta towards an AI strategy: high level policy document for public consultation. https://malta.ai/wp-content/uploads/2019/04/Draft_Policy_document_-_online_version.pdf. Accessed 12 June 2021
3. Government of Malta (2019c) Malta AI. https://malta.ai/. Accessed 10 Aug 2021
4. Government of Malta (2019d) Malta: the ultimate AI launchpad. https://malta.ai/wp-content/uploads/2019/11/Malta_The_Ultimate_AI_Launchpad_vFinal.pdf. Accessed 21 Aug 2021
5. Government of Malta (2019e) Malta:towards trustworthy AI: Malta's ethical AI framework. MDIA, s.l.
6. Government of Malta (2019f) Mapping tomorrow: a strategic plan for the digital transformation of the public administration 2019–2021. Office of the Principle Permanent Secretary (Office of the Prim Minister) and MITA, Valletta
7. Government of Malta (2019g) Mapping tomorrow; A strategic plan for the digital transformation fo the public admnistration 2019–2021. Government of Malta, Valletta
8. Snelson CL (2016) Qualitative and mixed methods social media research: a review of the literature. Int J Qual Methods 15(1):1–15
9. Almeida F (2018) Strategies to perform a mixed methods study. Eur J Educ Stud 5(1):137–151

Chapter 7
Research Philosophy—Pragmatic Paradigm

Research philosophy is a collection of ideas about the proper way to gather, assess and apply facts on a selected subject; as such the term is translated as the process of knowledge development [1]. Research is based on assumptions that unavoidably influence the interpretation of findings, which techniques are utilised and how the research question is perceived [2]. In this regard, what is considered to be true knowledge is defined as epistemological assumptions [3], while the factual contexts derived from the research are addressed as ontological assumptions [4]. Furthermore, axiological assumptions refer to the effect the researchers' stance has on the research process [5]. Thus, a strong research philosophy is founded on a set of well-considered and consistent assumptions that influence methodological choices, study design, data collection strategies, and analytic procedures. This facilitates the development of a coherent research study in which all components of the investigation are incorporated.

7.1 Human Nature and Ontology

Ontology is a set of assumptions regarding the relation of people, society and the world at large. The two main alternatives of ontology are subjectivism and objectivism. Subjectivity asserts that reality does not exist without people's perceptions of it, implying that reality is founded on each individual's unique experience and changes over time and circumstance [6]. Objectivity implies that reality as a fact is independent of the individuals who live it and their experiences [7].

M. Sciberras and A. Dingli, *Investigating AI Readiness in the Maltese Public Administration*, Lecture Notes in Networks and Systems 568,
https://doi.org/10.1007/978-3-031-19900-4_7

7.2 Epistemology

Epistemology is the accepted knowledge in a field of research. It is also reliant on how the researcher ascribes understanding to the study [8], which has three aspects: positivism regarding tangible knowledge domains or empirical research; realism which also involves empirical research but contains diverse researcher analyses, and interpretivism which applies the individuals' realty to the study [9].

In this study, the epistemological position is embedded in a mixed perspective of an ontological premise. The research recognises that knowledge is founded on the surrounding environment in which individuals partake, experience and survive. Knowledge is obtained by examining the nature of links between realities and comprehending the perception of humans in the context of the research question as a social reality. As a result, the interpretive perspective appears to be useful in establishing knowledge and the cause-and-effect correlation using primary sources. The researcher in this study suggests that several factors influence the adoption of AI solutions within public institutions.

7.3 Pragmatic Paradigm

In light of the research topic and context, a pragmatic paradigm is best suited as it enables the researcher to investigate and assess internal and external circumstances related to the research topic without constraints [10]. Furthermore, the pragmatic perspective is capable of explaining mixed-methods research since it assumes that there is an external environment that can only be shifted through human experiences [5]. The pragmatic paradigm's use is supported by the use of mixed methods in this study, the paucity of research on AI in Malta, and the possibility to integrate positivism and interpretivism in the same analysis.

References

1. Howell KE (2013) An introduction to the philosophy of methodology, 1st edn. SAGE Publications Inc., London
2. Wolgemuth JR, Hicks T, Agosto V (2017) Unpacking assumptions in research synthesis: a critical construct synthesis approach. Educ Res 46(3):131–139
3. Chamberlain K (2015) Epistemology and qualitative research. In: Rohleder P, Lyons AC (eds) Qualitative research in clinical and health pschology. Palgrave Macmillan, Hampshire, pp 9–28
4. Blaikie N (2018) Confounding issues related to determining sample size in qualitative research. Int J Soc Res Methodol 21(5):635–641
5. Biddle C, Schafft KA (2015) Axiology and anomaly in the pratice of mixed methods work: pragmatism, valuation and transformative paradigm. J Mixed Methods Res 9(4):320–334
6. Ragab MA, Arisha A (2018) Research methodology in business: a starter's guide. Manage Organ Stud 5(1):1–14

7. Oksuzoglu-Guven G (2016) Objectivity and research ethics in participant observation. Anthropologist 1(25):1–2
8. Dana LP, Dumez H (2015) Qualitaitive research revisited: epistemology of a comprehensive approach. Int J Entrep Small Bus 26(2):154–170
9. Al-Ababneh M (2020) Linking ontology, epistemology and research methodology. Sci Philos 8(1):75–91
10. Parvaiz GS, Mufti O, Wahab M (2016) Pragmatism for mixed methods research at higher education level. Bus Econ Rev 8(2):67–79

Chapter 8
Research Approach—Mixed-Methods Approach

A mixed-methods model is applied to this study to better address the research question. Mixed research combines quantitative and qualitative approaches in a single study to provide comprehensive understanding about a phenomenon [1]. Moreover, it can be integrated in such a manner that the original structures and procedures of qualitative and quantitative approaches are preserved [2]. In the social sciences, mixed-methods research has gained momentum as a form of enquiry that purposefully and methodically integrates qualitative and quantitative approaches to answer particular issues [3]. In this regard, data is collected, combined and analysed by the researcher and results are drawn from data collected through both approaches [4].

Through the application of the qualitative and quantitative approach, a triangulation design is facilitated, to compare the outcomes. For this study, the aftermath of the findings was also gauged vis-à-vis the *Malta: The Ultimate AI Launchpad—A National AI Strategy* document [5–11] and similar available literature. In view that AI processes within public administration in Malta is a new concept, it is thought that, by using the mixed approach, the researcher can gather an extensive collection of data from multiple sources operating at different levels. The process will maximise the feedback, hence allowing for justified and reliable recommendations.

References

1. Timans R, Wouters P, Heilbron J (2019) Mixed methods research: what it is and what it could be. Theory Soc 48(2):193–216
2. Wilson V (2016) Research methods: mixed methods research. Evid Based Libr Inf Pract 11(1 (S)):56–59
3. Meixner C, Hathcoat JD (2019) Tha nature of mixed methods research. In: Liamputtong P (ed) Handbook of research methods in health social sciences. Springer, Singapore, pp 51–70
4. Tashakkori A, Teddlie C (2010) SAGE handbook of mixed methods in social & behavioral research, 2nd edn. SAGE Publications Inc., London

M. Sciberras and A. Dingli, *Investigating AI Readiness in the Maltese Public Administration*, Lecture Notes in Networks and Systems 568,
https://doi.org/10.1007/978-3-031-19900-4_8

5. Government of Malta (2019a) MALTA the utimate AI launchpad; A strategy and vision for artifical intelligence in Malta 2030. Government of Malta, Valletta
6. Government of Malta (2019b) Malta towards an AI strategy: high level policy document for public consultation. https://malta.ai/wp-content/uploads/2019/04/Draft_Policy_document_-_online_version.pdf. Accessed 12 June 2021
7. Government of Malta (2019c) Malta AI. https://malta.ai/. Accessed 10 Aug 2021
8. Government of Malta (2019d) Malta: the ultimate AI launchpad. https://malta.ai/wp-content/uploads/2019/11/Malta_The_Ultimate_AI_Launchpad_vFinal.pdf. Accessed 21 Aug 2021
9. Government of Malta (2019e) Malta:towards trustworthy AI: Malta's ethical AI framework. MDIA, s.l.
10. Government of Malta (2019f) Mapping tomorrow: a strategic plan for the digital transformation of the public administration 2019–2021. Office of the Principle Permanent Secretary (Office of the Prim Minister) and MITA, Valletta
11. Government of Malta (2019g) Mapping tomorrow; A strategic plan for the digital transformation fo the public admnistration 2019–2021. Government of Malta, Valletta

Chapter 9
Research Analysis—Triangulation Approach

Triangulation is a research approach that combines numerous data sources, theories or research methodologies to guarantee that a research study's data, analysis, and findings are as thorough and precise as possible. It also aids in the identification of topics that require additional investigation [1]. This study applied methodological triangulation and data triangulation. Methodological triangulation regards the utilisation of qualitative and/or quantitative methodologies to study the phenomenon and involves findings from different data collection methods to be compared [2] to determine whether they provide comparable results. Validity is established when the findings from each technique are the same [3]. Diversely, data triangulation refers to the use of several sources of data to improve the credibility of the research. However, each technique may also focus on a distinct element of the underlying phenomena, resulting in conclusions that are either complementary or divergent to one another [4, 5].

As the central research question being addressed is to investigate the extent of AI readiness within the Maltese public administration, and considering there is no local literature about the domain, it was deemed necessary to conduct this research to gain insight into the internal perceptions and impressions on AI as part of the everyday working processes of the public workforce. Henceforth, the triangulation approach to data analysis cross-validated the qualitative results to the quantitative outcome, which findings substantiated the final proposals. Figure 9.1 illustrates the triangulation design applied for this research.

M. Sciberras and A. Dingli, *Investigating AI Readiness in the Maltese Public Administration*, Lecture Notes in Networks and Systems 568,
https://doi.org/10.1007/978-3-031-19900-4_9

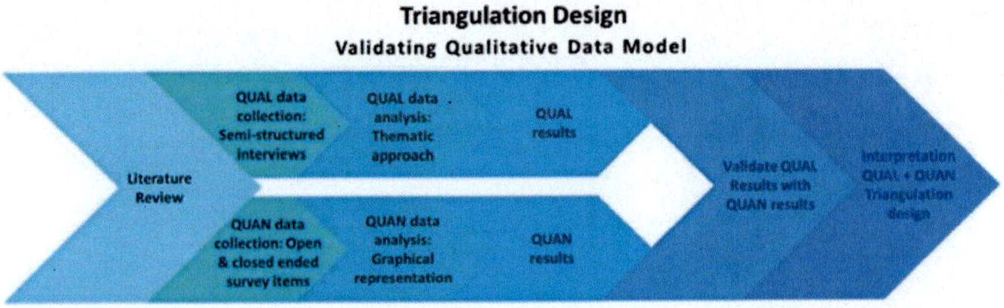

Fig. 9.1 Triangulation design: validating qualitative data model. *Source* Adopted from [6]

References

1. Moon MD (2019) Triangulation: a method to increase validity, reliability, and legitimation in clinical research. J Emerg Nurs 45(1):103–105
2. Bekhet AK, Zauszniewski JA (2012) Methodological triangulation: an approach to understanding data. Nurs Res 20(2):40–43
3. Farquhar J, Michels N (2016) Triangulation without tears. In: Groza M, Ragland C (eds) Marketing challenges in a turbulent business environment. Springer, Cham, pp 325–330
4. Almeida F (2018) Strategies to perform a mixed methods study. Eur J Educ Stud 5(1):137–151
5. McCarthy JR, Holland J, Gillies V (2003) Multiple perspectives on the 'family lives' of young people: methodological and theoretical issues in case study research. Int J Soc Res Methodol 6(1):1–23
6. Creswell J, Plano CV (2007) Designing and conducting mixed methods research. Sage, Thousand Oaks, CA

Chapter 10
Qualitative Research and Findings

Aspers and Corte [1] define qualitative research as a continuous practice whereby the researchers gain a better understanding of the subject being investigated, by identifying new important distinctions or enforcing relationships. As a result, applying a qualitative study as part of the research study ensures that primarily it makes up for the lack of information available, and also collects the necessary information from primary sources.

10.1 Data Collection Instrument: Semi-structured Interviews

Data collection for the qualitative research was carried out via the application of semi-structured interviews. Semi-structured interviews are a versatile data-collecting method because they provide greater depth by allowing the interviewer to explore and expand on the interviewee's replies [2]. The interviews were conducted over video conferencing and audio-recorded for ease of reference and transcription purposes.

10.2 Design of Interview Questions

The interview questions stemmed from the literature available, which discusses the factors that allow for AI readiness in both public and private organisations. Refer to Appendix B for a copy of the interview questions.

M. Sciberras and A. Dingli, *Investigating AI Readiness in the Maltese Public Administration*, Lecture Notes in Networks and Systems 568,
https://doi.org/10.1007/978-3-031-19900-4_10

10.3 Target Population—Purposive

The target population refers to the identified group of people who were chosen based on inclusion and exclusion criteria related to the research topic [3]. A purposive sampling approach was applied to this study. The purposive method is a non-probability sampling approach in which the individuals partaking the research are selected based on the researcher's judgment. By adopting the purposive approach, it was assumed that information gathering would be valid and related to the study.

Consequently, Chief Information Officers (CIOs) representing the Government of Malta and Information and Communication Technologists (ICTs) were selected to participate in the interviews; the CIOs being the individuals implementing AI within the ministries and the ICTs representing the Maltese public sector, as well as assisting in AI adoption within the governmental institutions and the private sector.

10.4 Sampling—Total Population

As for the sampling approach, it was decided to opt for the total population of CIOs within the public service and a selection of experts from the public sector. In total, 14 CIOs and 4 ICTs took part in the research study. As the selected interviewees are considered experts in the field of AI implementation, this method allowed for gaining a better perspective and detailed insight on what the participants deem as facilitating AI adoption, and those factors that are lacking, hence hindering AI readiness within the public administration. In this regard, expert sampling was a useful technique to employ to determine also whether or not additional investigation is worthwhile [4].

10.5 Data Analysis—Inductive Thematic Analysis

Qualitative should demonstrate that data analysis was performed precisely, consistently, and exhaustively by documenting, systematizing, and disclosing the procedures of analysis in sufficient detail to allow the reader to determine the process's credibility [5].

In research, the inductive technique entails making actual observations about a particular research subject and constructing ideologies and possibilities based on the study [6]. Thereupon, the research underpinned an inductive approach, whereby observations were gathered through semi-structured interviews, which resulted in a thematic analysis. To this extent, inductive thematic analysis is described as the process of coding data without fitting it into a pre-defined coding framework or the researcher's diagnostic anticipations. The themes generated via an inductive method

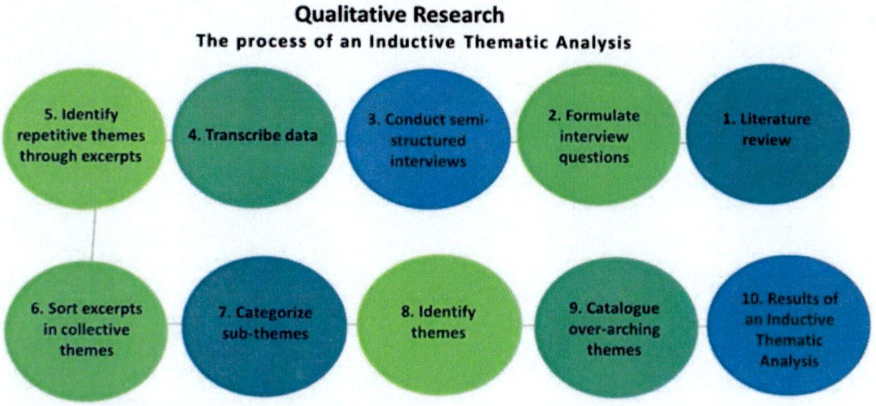

Fig. 10.1 Qualitative research: the process of the inductive thematic analysis. *Source* Author's representation of this study's qualitative research

are closely related to the data and may have little relevance to the particular questions asked of the participants [7]. Hence this type of thematic analysis, which was solely data-driven, was deemed appropriate to this research. Figure 10.1 indicates the qualitative process undertaking for this study.

10.6 Research Credibility and Trustworthiness—TACT Framework

When conducting research, techniques for guaranteeing rigour should be integrated into the process from the initial stages [8]; however, due to the diversity and creativity applied when researching qualitatively, there are no widely recognised standards for assessing validity or reliability [9]. In this regard, this investigation adopts the TACT framework that allows for assessing the rigour of qualitative research in four dimensions: trustworthiness, auditability, credibility and transferability [10].

To attain trustworthiness, it is important to use a methodical approach in examining and categorising data [11]. The study applied a rigorous approach whereby data was read multiple times to ensure the maximum extraction of excerpts, from which sub-themes emerged. These sub-themes were then converged in themes and over-arching themes, as illustrated in Table 10.1.

Auditing is the method of carefully examining the judgments and choices made throughout qualitative research. The word auditability refers to the level of clarity the research provides for it to be assessed, calling for an audit trail of evidence that runs from the beginning to the end of the research, making it easier to comprehend the decisions made [12]. The investigation's step-by-step process is detailed in Appendix D.

Table 10.1 Qualitative research: overarching themes, themes and sub-themes. *Source* Author's representation of qualitative research

Overarching themes		Themes		Sub-themes	
1	Understanding AI	1.1	AI as a tool	1.1.1	Powered by data
				1.1.2	Assistive to humans
				1.1.3	Autonomous
				1.1.4	Intelligent
		1.2	Current uses of AI	1.2.1	Cloud
				1.2.2	Big data
				1.2.3	Semantic analysis
				1.2.4	Chatbots and NLP
				1.2.5	Different AI applications
2	Policy and procedures	2.1	National AI strategy	2.1.1	Pros and Cons
		2.2	Ministry's own	2.2.1	Strategic vision
				2.2.2	Staying focused
				2.2.3	No ministerial strategy
				2.2.4	Inhouse strategy—evidence based
3	Data	3.1	Availability	3.1.1	Data collection
				3.1.2	Data structure
				3.1.3	Data dumping
				3.1.4	Data quality
		3.2	Data processing	3.2.1	Legal frameworks
				3.2.2	Ethical AI
4	Preparedness	4.1	Ready or not	4.1.1	Not ready
				4.1.2	Getting there
				4.1.3	Ready
		4.2	Technicalities		
		4.3	Availability of funds		
		4.4	Competitivity		
5	Change what?	5.1	Current practices	5.1.1	Set in old ways
				5.1.2	Institutional structure
				5.1.3	Diffuse misconceptions
				5.1.4	Increase awareness
				5.1.5	Dissolve silos
		5.2	To drive AI	5.2.1	Positivity prerequisite
				5.2.2	Leadership

(continued)

Table 10.1 (continued)

Overarching themes		Themes		Sub-themes	
				5.2.3	Public inclusion
6	In demand	6.1	Knowledge and expertise	6.1.1	Lack of it
				6.1.2	ICT class
				6.1.3	Role of the CIO
				6.1.4	Outsourcing
		6.2	Educational movement	6.2.1	Job retention
				6.2.2	Career paths
				6.2.3	Agencies to the rescue
				6.2.4	Mainstream education
7	AI as an asset	7.1	Enhancing processes		
		7.2	Effective use of resources		
		7.3	Evidence based decisions		
		7.4	Investment		
		7.5	Challenges		
8	Addressing risks	8.1	Misuse of funds		
		8.2	Mal-intent of AI solutions		
		8.3	Risk mitigation		
		8.4	Identify accountability		
		8.5	Data quality		
		8.6	Discontinuity of activity		
9	Future applications of AI	9.1	Citizen centric		
		9.2	Image processing		
		9.3	Analysis		
10	Moving forward	10.1	Needs analysis		
		10.2	Informative campaign		
		10.3	More to be done	10.3.1	Addressing the nation
				10.3.2	Within the public administration
				10.3.3	Increased collaboration
		10.4	Tread cautiously		

Credibility determines whether or not a study is valid and reliable by thoroughly examining the data, data analysis, and findings. As such, the credibility of this investigation was established through the use of data (literature, interviews and surveys) and source (expertise and public administration workforce) triangulation technique, ensuring the integrity of the data collected [13].

In qualitative research, transferability means that the findings of a study may be transferred to other contexts and it suggests that such outcomes can teach useful lessons to other comparable situations. Additionally, the integrity of the research should be examined through the evidence provided [14], in this case, the excerpts representing the interview participants and the graphical interpretations of the survey. Furthermore, to ensure transferability, experts of the phenomenon should form the participants' sample [15], which is also in line with the recruitment of CIOs and ICT experts for the qualitative study.

Hence, to address the TACT framework, the following procedures have been carried out:

- Expert participants were selected to partake in the research study. Ethical considerations were practised throughout the investigation.
- Transcription of the interviews was manually done by the researcher. This allowed the researcher to connect with the data presented, as well as immediately identify emergent themes.
- The researcher was able to mark any major changes in the participants' tone of voice using high-quality recording equipment. This enabled a deeper understanding of the participants' views about AI readiness in their Ministry.
- An audit trail was created, documenting all of the research steps done from the beginning to the completion of the investigation as per Appendix D.
- To assure truthfulness of data, transcripts were returned to the participants to verify the accuracy of text before data analysis.
- Upon approval, maximum excerpts were identified, which excerpts were then sorted into collective themes as per Appendices E–N.

10.7 Qualitative Research Findings

Ten overarching themes have been identified from the interviews that were held with the 18 participants, and they encapsulate an overarching notion that underpins a variety of themes. The identified 32 themes provide further detail through 42 identified sub-themes which focus on an important factor of the original concept. Table 10.1 illustrates the overarching themes, themes and sub-themes, and each is linked to one another.

Overarching theme 1: Understanding AI is divided in two themes, one being AI application as a tool, and the other mentioning ongoing AI practices within our ministries. As such, the sub-themes indicate the initial knowledge of the concept of AI from different viewpoints and how AI is applied in these terms.

Overarching theme 2: Policy and Procedures investigates the application of strategies and internal policies to address the introduction of AI in public administration. In this regard, a number of participants currently seem to lack a clear vision on the role of AI within their ministry while others are focused on the end goals and how implementing AI in their remit will transform results. The theme of national AI strategy also identifies the advantages and disadvantages of having a national strategy in place.

Overarching theme 3: Data is the foundation and building blocks of AI processes and solution, hence its importance is highlighted throughout the research. Data availability is further dissected into data collection methods, data dumping and structures in place, as well as the importance of data quality. Data processing looks into the legal frameworks and ethical AI in various scenarios.

Overarching theme 4: Preparedness captures the readiness stages of the public service, as not yet ready, in the process of getting ready and ready to implement. Technical matters are captured from different perspectives, while funding approaches explore budgeting practices. The sub-theme addressing competitivity discusses the controversy of financial flexibility within the public administration.

Overarching theme 5: Change What? encapsulates the foreseen changes needed to implement AI as part of the everyday working solution. This is done by addressing and investigating the public service mentality, and revising the institutional structures and whether they are effective to the current work practices in the public service. Diffusing misconceptions about AI among the public service as well as increasing awareness about AI concepts would ensure acceptance of emerging technologies. The urgent need to eliminate silos was discussed by all participants; this will increase collaboration and data sharing in many aspects of implementation.

Overarching theme 6: In Demand examines the true need for knowledge and expertise within the public administration and how the lack of these components is affecting the ICT class, which is directed by the CIO. In this regard, outsourcing seems to be a necessity in order to implement novel solutions, which factor is automatically linked to the need of an educational movement in order to assure job retention, the need for career paths within the public administration, how agencies are facilitating the exchange of AI knowledge and the role of mainstream education in raising a knowledgeable and responsible AI-friendly society.

Overarching theme 7: AI as an Asset presents the benefits the ministries already implementing AI are experiencing, as well as the envisaged advantages that an AI system will grant to its users. Enhancement of workflows, using both human and financial resources effectively, and the prediction of possible outcomes are set to aid public servants in their day-to-day administration. AI as an Asset also explores the possibility of investment coming to Malta, as well as the challenges, also related to other themes, that the public administration is anticipating.

Overarching theme 8: Addressing Risks concerns matters that might arise if AI applications are misused, if funds are wasted and if data is of an inferior quality.

Additionally, identifying accountability and keeping decision-making in the hands of humans seems to be an absolute necessity when adopting predictive and autonomous AI systems. The termination of AI activities was also discussed.

Overarching theme 9: Future Applications of AI seem to be citizen-centric and focused on enhanced business services via image processing and several analysis solutions that would assist the public service in delivering an enhanced service.

Overarching theme 10: Moving Forward discusses different viewpoints of the participants in terms of a needs analysis that is necessary to identify which processes can benefit from the adoption of AI in the short and long term, and how an informative campaign will empower citizens and public servants with the knowledge needed to trust AI. The theme More to be Done is divided in three sub-themes; the Addressing the Nation and Within the Public Administration themes address the importance of education to facilitate AI adoption, while the third theme—Increased Collaboration between private sector, public administration and academics—can foster exploration of new projects while also investigating the best practices of AI solutions. Taking caution in adopting and promoting AI was also discussed in terms or ensuring that Malta reaps the best benefits in the long term, rather than applying exaggeration in implementing AI-enhanced methods.

References

1. Aspers P, Corte U (2019) What is qualitative in qualitative research. Qual Sociol 42(2):139–160
2. Adams WC (2015) Conducting semi-structured interviews. In: Wholey JS, Hatry HP, Newcomer KE (eds) Handbook of practical program evaluation. Wiley & Sons Inc, CA, pp 492–505
3. Asiamah N, Mensah HK, Oteng-Abayie EF (2017) General, target and accessible population: demystifying the concepts for effective sampling. Qual Rep 22(6):1607–1622
4. Etikan I, Musa SA, Alkassim RS (2016) Comparison of convenience sampling and purposive sampling. Am J Theor Appl Stat 5(1):1–4
5. Nowell LS, Norris JM, White DE, Moules NJ (2017) Thematic analysis: striving to meet the trustworthiness criteria. Int J Qual Methods 16:1–13
6. Woiceshyn J, Daellenbach U (2018) Evaluating inductive vs deductive research in management studies: implications for authors, editors and reviewers. Qual Research Organ Manage Int J 13(2):183–195
7. Hanafizadeh P, Harati Nik MR (2020) Configuration of data monetization: a review of literature with thematic analysis. Glob J Flex Syst Manag 21:17–34
8. Cypress B (2017) Rigour or reliability and validity in qualitative research: perspectives, strategies, reconceptualization and recommendations. Dimens Crit Care Nurs 36(4):253–263
9. Hayashi PJ, Abib G, Hoppen N (2019) Validity in qualitative research: a processual approach. Qual Rep 24(1):98–112
10. Daniel BK (2019) What constitutes a good qualitative research study? Fundamental dimensions and indicators of rigour in qualitative reseach: the TACT Framework. Academic conferences international limited, Kidmore End
11. Creswell JW, Miller DL (2000) Determining validity in qualitaitve inquiry. Theory Pract 39(3):124–130
12. Nair LB (2020) From 'Whodunit' to How@: detective stories and auditability in qualitative business ethics research. J Bus Ethics 1:1–15

13. Stacey A (2019) Proceedings of the 18th European conference on research methodology for business and management studies, wits business school, Johannesburg, South Africa, 20–21 June 2019. Academic conference and publishing intenational ltd., Reading
14. Cope DG (2014) Methods and meanings: credibility and trustworthiness of qualitative research. Oncol Nurs Forum 40(1):89–91
15. Forero R et al (2018) Application of four dimension criteria to assess the rigour of qualitative research in emergency medicine. BMC Health Serv Res 18(1):1–11

Chapter 11
Quantitative Research

Quantitative research is a collection of approaches for conducting systematic studies of social issues using numerical data; thus, the approach requires the quantification and measurement of the subject being assessed [1]. This inquiry adopts a descriptive data-driven approach as through the survey, the participants describe how ready they feel about the adoption of AI and also form an opinion on how they think AI should be introduced. Moreover, through a descriptive model, the researcher formed a theory once the data was analysed, hence the data-driven approach to results.

11.1 Data Collection Tool: Survey

Data collection for the quantitative research was performed through the dissemination of a questionnaire. The survey included a selection of open and close-ended questions. As the survey was intended for a portion of the population of which the researcher was uncertain how informed they were on AI, the close-ended questions were presented on a Likert Scale to provide structure and minimal information on what the adoption of AI could entail. The survey was designed using the SurveyMonkey website.

11.2 Design of Survey Questions

The survey questions emanated from the findings derived through the qualitative research, with the purpose being to assess whether the experts and their workforce fostered similar impressions or otherwise. The survey findings were used as a measuring tool to the qualitative findings.

M. Sciberras and A. Dingli, *Investigating AI Readiness in the Maltese Public Administration*, Lecture Notes in Networks and Systems 568,
https://doi.org/10.1007/978-3-031-19900-4_11

11.3 Target Population—Maltese Public Administration

The Maltese public administration was selected as the target audience as this labour force is in the frontline of experiencing the change that AI adoption will bring about.

When researching public administration, public service and public sector, the terms are intertwined and most often lead to the same meaning. The Maltese public administration refers to the management of public affairs [2] through the employees operating under the Public Service and the public sector. In the Maltese jurisdiction, these two (2) are distinguished as defined by the public service authorities [3]. The Public Service constitutes of the ministries and Government departments of which employees are directly engaged with the government and operate on the governments' payroll. These public officers are categorised in grades and salary scales as directed by the Office of the Prime Minister (OPM) and can move in between ministries to develop their careers. On the other hand, the public sector incorporates the Governments' agencies, foundations and authorities, all of which provide a public service however, operate autonomously from the government and hold an independent legal capacity.

Hence, the public administration in Malta is identified as:

i. The public service—which is the personnel engaged by ministries and on the ministries' payroll. These civil servants are categorised in scales and grades as per the government's schedule: scale 1–4 are top management, scale 5–7 are senior management, scale 8 and 9 are middle management and from scale 10 to scale 20 various operational duties [4]
ii. The public sector—involves the employees of the government's foundations, agencies and authorities. These entities, although they identify with a particular ministry, operate autonomously from the ministries [5–7].

11.4 Sample—Total Population

The total population of the public administration as issued by the NSO totals 50,808 employees [8]. In order to maximise the potential number of participants, the total population of public administrators was invited to partake the survey. This choice of non-probability sampling allowed for a combination of replies targeting multiple variables.

11.5 Data Analysis—Graphical Interpretation

In order to translate the data statistically, the researcher used the Microsoft Excel program. This facilitated a descriptive data interpretation.

11.6 Variables

In quantitative research, the term variables refer to data that is measured in numbers. The variables assigned to this research include the age of the participants, the scale on which their employment is assigned and the years they have been public employees.

11.7 Reliability and Validity

While the terms validity and reliability are often used synonymously, they relate to distinct characteristics of a measurement technique. [9]. Whiston [10] defined validity as the acquisition of data that is fit for the research topic through the selected measuring tool. Hence, the correct interpretation of information from the survey is essential to establish the validity of the research outcome. This investigation adopts a descriptive validity whereby respondents' data is factually reported through statistical measurement.

The extent to which an identical test produces a comparable result when used numerous times is referred to as reliability, hence referring to the measurement's consistency [11]. In this regard, the researcher allowed for a 5% margin of error and 95% confidence level when surveying to draw the minimum number of responses necessary for the survey to be considered reliable. As the population targeted for the survey totaled 50,808, the survey reliability resulted in a minimum of 381 survey replies. Hereof, as a total of 494 participants partook in the survey, quantitative research reliability was assured. For the purpose of the research, the standard quantitative research formula was applied.

11.8 Quantitative Research Findings

A survey was distributed among the public administration in order to gain feedback and insight into public servants' perspectives about adopting AI as part of their workflows. The questionnaire was disseminated online via the ministries' own channels and 494 respondents took part in the ten open and close-ended questions. The Likert scale was also applied to several questions in order to grasp the level of AI knowledge among the workforce and how they would like AI to be adopted. The following is an illustrative interpretation of the results for each question. Microsoft Excel was applied to generate a graphical version of the replies received.

Demographics

i. **Age**

A total of 139 participants were aged between 45 and 54, and 131 participants fell within the bracket of 35–44 years. Another 120 respondents were aged between 25

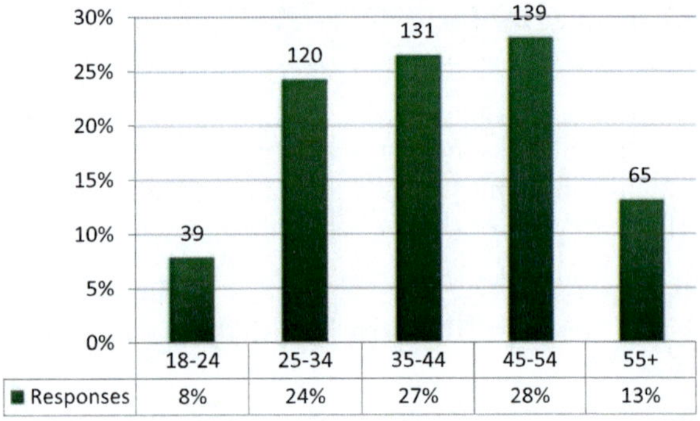

Fig. 11.1 Age. *Source* Author's representation of quantitative results

and 34, while 65 replies derived from participants aged over 55. There were 39 replies from the youngest age group (Fig. 11.1).

ii. **Gender**

A total of 260 females and 232 males took part in the survey, while 2 participants identified as other (Fig. 11.2).

iii. **Salary Scale**

As reflected in Fig. 11.3, the majority of participants fell in scale 7 and scale 5, scoring 80 and 75 respectively. There were 54 respondents registered at scale 10, and another 50 at scale 6. In addition to this, 38 participants fell in scale 9, and there were 35 participants in scale 8 and scale 11. From scale 4, there were 29 respondents, and scale 3 registered 17 replies. In total, 41 replies derived from scales 12, 13 and 14, while 22 in total were registered under scales 15 and 16. A total of 13 respondents that partook the survey fell in scales 1 and 2, while the remaining 5 replies derived from scales 18, 19 and 20.

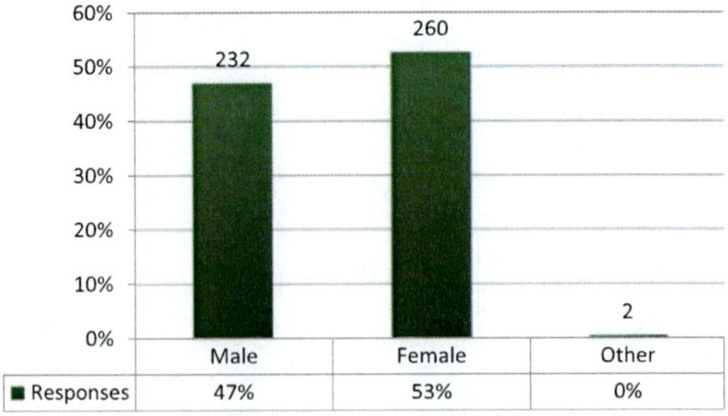

Fig. 11.2 Gender. *Source* Author's representation of quantitative results

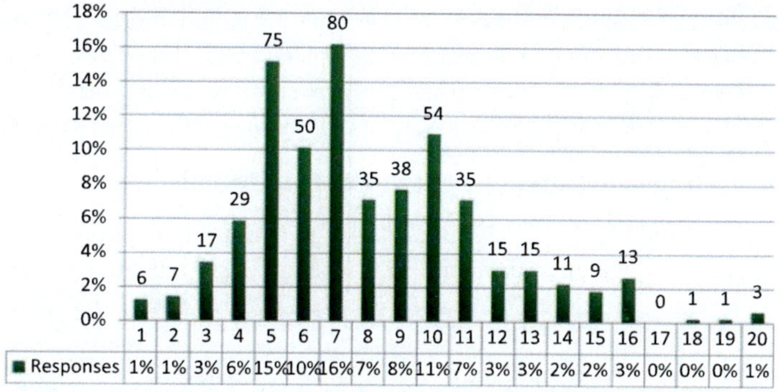

Fig. 11.3 Salary scale demographics. *Source* Author's representation of quantitative results

iv. Years of Employment in Public Administration

In terms of experience, 152 respondents fell in the bracket of 0–5 years as employees within the public administration. Next, 71 participants had been engaged as public servants between 6 and 10 years, while 65 registered 11–15 years of experience. Within the brackets of 16–20 and 21–25 years fell 36 and 37 respondents respectively. Additionally, 55 participants had 26–30 years as registered employees, and 51 participants registered 31–35 years as public servants. The remaining figures were distributed as follows: 15 participants had been public servants between 36 and 40 years, 11 respondents had between 41 and 45 years, and 1 participant had over 46 registered employment years within the public service (Fig. 11.4).

Q1. What is Artificial Intelligence (AI)?

Question 1 led to a close-ended preference or an open-ended reply.

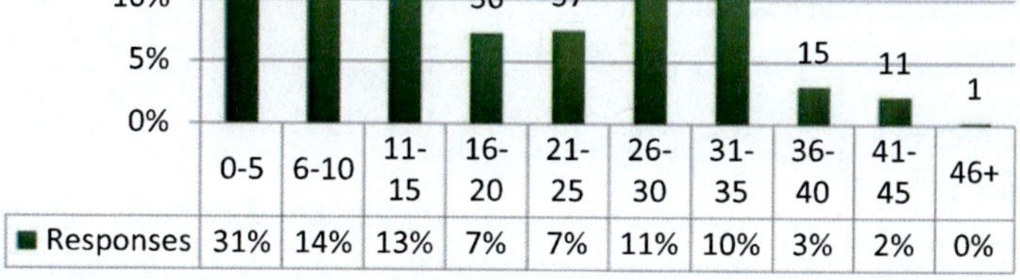

Fig. 11.4 Years employed in public administration. *Source* Author's representation of quantitative results

In fact, 124 respondents were not sure of what AI is and another 12 respondents ticked the "do not know" option as per Fig. 11.5. However, 358 participants provided an answer to what AI is in their terms. A thematic table in this regard illustrates their response, of which one reply could cover several themes, as presented in Table 11.1.

Q2. Are AI technologies already in use in your ministry?

Question 2 presented close-ended replies or the possibility to explain how AI is used in the respondents' respective ministry. In this regard, 262 respondents were unsure whether AI technologies were already available and 162 stated that there weren't any AI solutions in place in their ministry (Fig. 11.6).

On the other hand, 70 participants provided a reply in relation to the technologies already implemented by their ministry. The answers were gathered in a thematic Table 11.2.

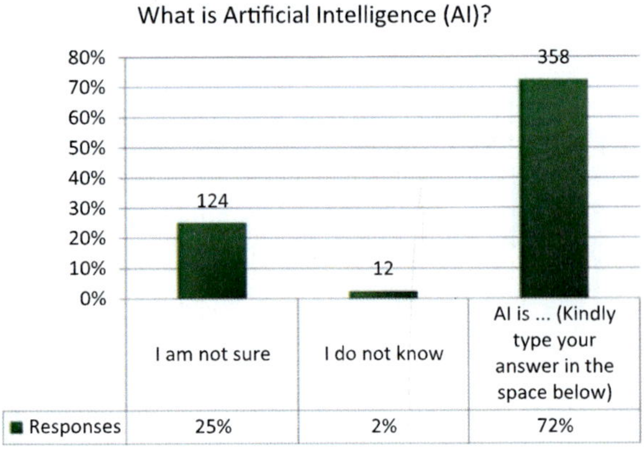

Fig. 11.5 What is artificial intelligence? *Source* Author's representation of quantitative results

Table 11.1 Understanding the term artificial intelligence. *Source* Author's representation of quantitative results

Theme	Number of times mentioned
Intelligence with the capacity of human beings	132
Robotic or machine intelligence	92
Use of data for accurate predictions and performing intelligent tasks	68
Intelligent computer programs that learn to take decisions and acts accordingly	60
Machine learning	57
Intelligent technology that facilitates work processes and improves quality of life	54
Smart technologies	36
Algorithms generating information and mimicking humans	29
Performing tasks without the need for humans	15

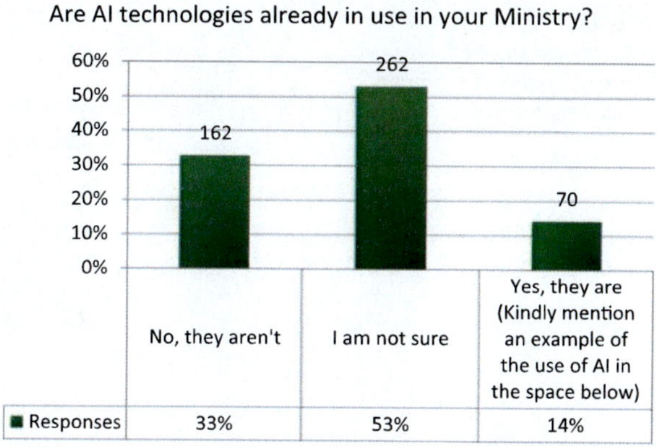

Fig. 11.6 Are AI technologies already available in your ministry? *Source* Author's representation of quantitative results

Table 11.2 AI technology in ministries as identified by the survey participants. *Source* Author's representation of quantitative results

Theme	Number of times mentioned
Medical field	23
Agriculture area	10
Computers/servers/laptops	8
Online systems and processing of applications	7
Smart parking	4
112 system	3
Microsoft cloud	3
Chatbots	2
HUB/NFC	1
Dakar/SRS/HRIMS	1
Fisheries	1
Drones	1
Epps	1
e-translation	1
Maltese spell checker	1
Proof of concept	1
Robotic arm	1
Satellite tools	1

Q3. Are you aware that in 2019, the Maltese Government launched a strategy and vision for AI adoption?

Question 3 resulted in two replies: not being aware of the National AI strategy and being aware of it. If the respondent was aware, they were invited to indicate whether they were aware whether the strategy will affect their role (Fig. 11.7).

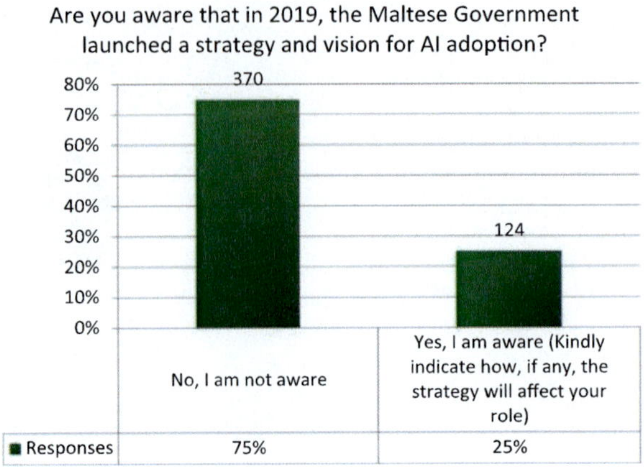

Fig. 11.7 Awareness of the AI national strategy. *Source* Author's representation of quantitative results

Through question 3, 370 participants replied that they were not aware of the National AI strategy which was launched in 2019. The remaining 124 participants who confirmed their knowledge about the strategy were asked to comment on whether they know if this will affect their role. The answers were gathered in themes presented in Table 11.3.

Table 11.3 Will the AI National strategy impact your role? *Source* Author's representation of quantitative results

Theme	Number of times mentioned
Increased workflow efficiency	34
Not aware if/how it will affect my role	22
The strategy does not affect my role	12
Heard through the media, not mentioned again	10
Assist in data manipulation and analytics	8
AI is part of my role	7
My role will be affected, not sure how	6
Projects which involve AI	5
Change management, policy measures designed to equip, reskill and upskill workers in every sector	4
Improved technology	4
Need more information about AI	4
Remote working	3
My role could be replaced by AI	2
Re-engineering of processes	1
I doubt its progress and implementation	1
Despite the benefits, there is hesitancy to use AI at my work	1
AI technology did not reach my section	1

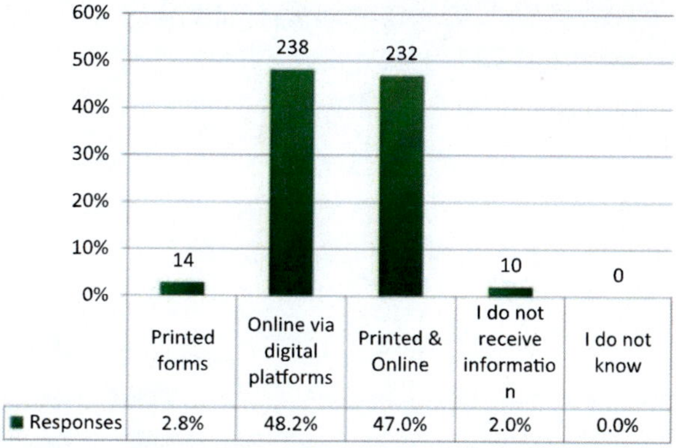

Fig. 11.8 As part of your role, how do you receive information? *Source* Author's representation of quantitative results

Q4. Data Collection, Analysis and Processing

Question 4 is divided in three parts that explore data collection and preferences.

Q4.1 As part of your role, how do you receive information?

According to the answers to question 4.1, 238 participants received information via online digital platforms while 232 received a mix of printed documents and through online systems. 14 replied that information is received in printed forms, while 10 did not receive information. Further to question 4.1, the respondents were asked to indicate on a Likert scale (Q4.2) how the data collection method affected their day-to-day role (Fig. 11.8).

Q4.2 By collecting information this way, I:

Q4.2 presents how the respondents feel about the data collection processes they practice at work. The three options and figures are illustrated as follows:

Q4.2.1 Find it difficult to do my job

A total of 347 participants were in consensus and pleased with the way data is collected, while 39 respondents agreed that the data collection methods practiced in their ministry was challenging in terms of following through their job requirements. Another 69 participants remained undecided, and 39 marked the question as not applicable (Fig. 11.9).

Q4.2.2 Slow down as it is time-consuming

In Fig. 11.10, 304 employees disagreed or strongly disagreed that the way data is collected slows down their workflows or is more time-consuming, while 79 agreed and 17 respondents strongly agreed that the data collection process is time-consuming and does slow them down. There were 71 participants who were undecided, and 23 ticked the not applicable option.

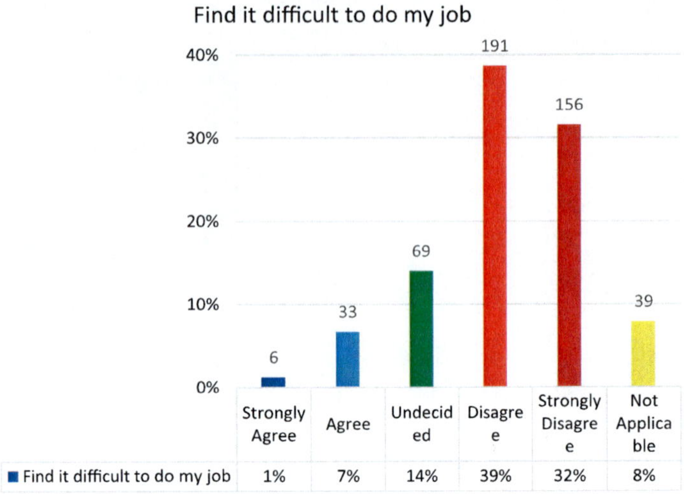

Fig. 11.9 By collecting information this way, I find it difficult to do my job. *Source* Author's representation of quantitative results

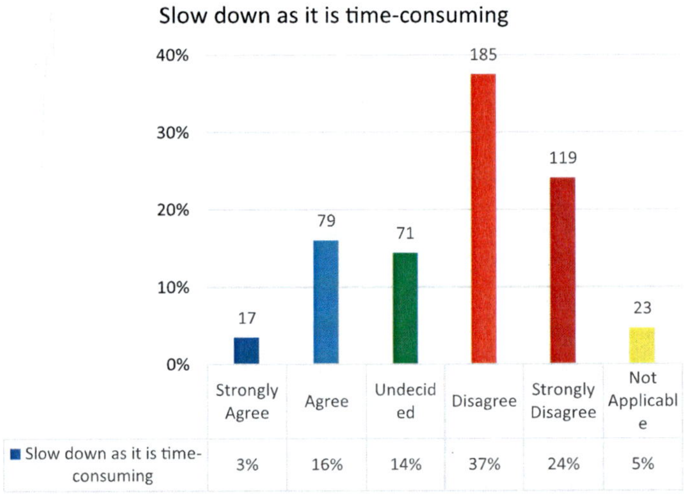

Fig. 11.10 By collecting information this way, I slow down as it is time consuming. *Source* Author's representation of quantitative results

Q4.2.3 Do my job faster

A total of 398 employees agreed that the way data is collected helps them to deliver their tasks faster, while 42 respondents confirmed that their data collection methods hindered their job performance. There were 43 and 11 participants who marked undecided and not applicable respectively (Fig. 11.11).

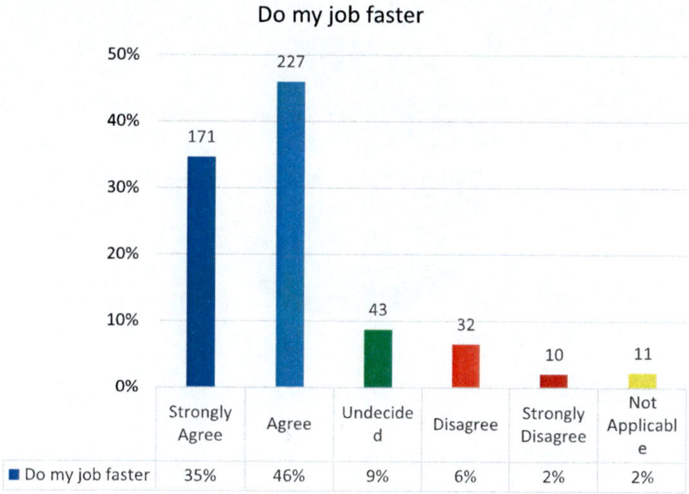

Fig. 11.11 By collecting information this way, I do my job faster. *Source* Author's representation of quantitative results

Fig. 11.12 Prefer information received on paper. *Source* Author's representation of quantitative results

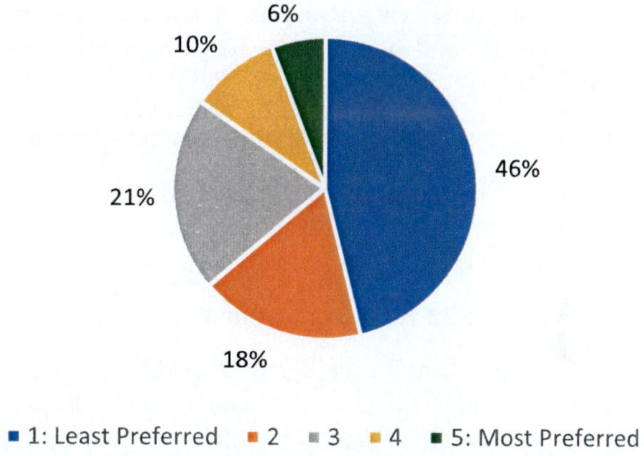

Q4.3 How do you prefer to receive information

Question 4.3 presented two options, as either information received on paper or digitally by choosing their least preferred as 1 and most preferred as 5. The replies are presented accordingly (Fig. 11.12).

A total of 228 employees confirmed that information received on paper was the least favourite; this is translated in Fig. 11.13, which illustrates that 327 participants preferred to receive information digitally.

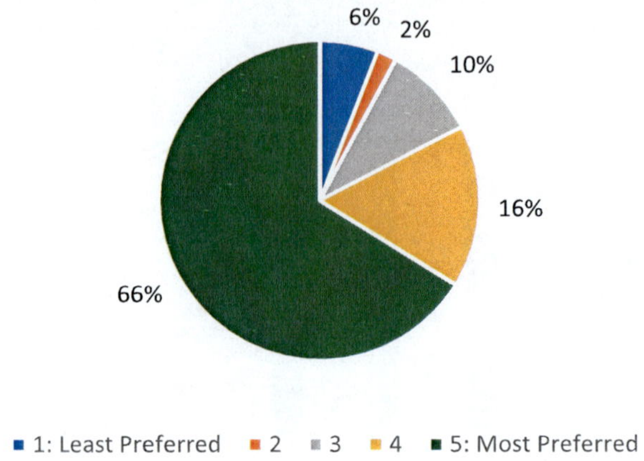

Fig. 11.13 Prefer information received digitally. *Source* Author's representation of quantitative results

Q5. AI at Work

Question 5 investigated the perception of the employees vis-a-vis AI at work. The respondents' feedback was captured in ten statements that were backed by literature. The statements and the respective reaction to the statement were presented on a Likert scale as follows:

Q5.1 AI can help me simplify and process information quicker

A total of 396 employees agreed that AI can help them simplify and process information quicker, while 15 respondents disagreed with the statement. There were 62 participants who were undecided and 21 who did not know (Fig. 11.14).

Q5.2 AI can help me solve problems

A total of 333 employees agreed that AI can help them solve problems at work. Another 114 participants were undecided about this statement and 24 respondents disagreed, while 23 reported that they did not know whether AI could help them at work (Fig. 11.15).

Q5.3 AI can help me make informed decisions

Regarding the statement that AI can help employees make informed decisions, 123 participants strongly agreed, 212 agreed, 24 disagreed, while 116 participants were undecided. There were 18 employees who did not know whether AI can help them make informed decisions (Fig. 11.16).

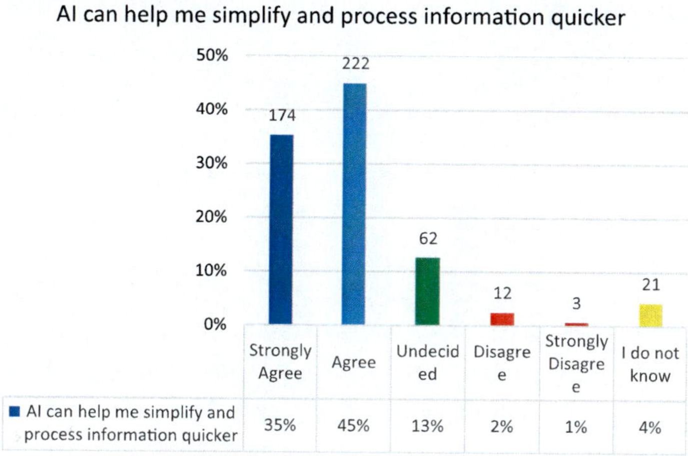

Fig. 11.14 At work, AI can help me simplify and process information quicker. *Source* Author's representation of quantitative results

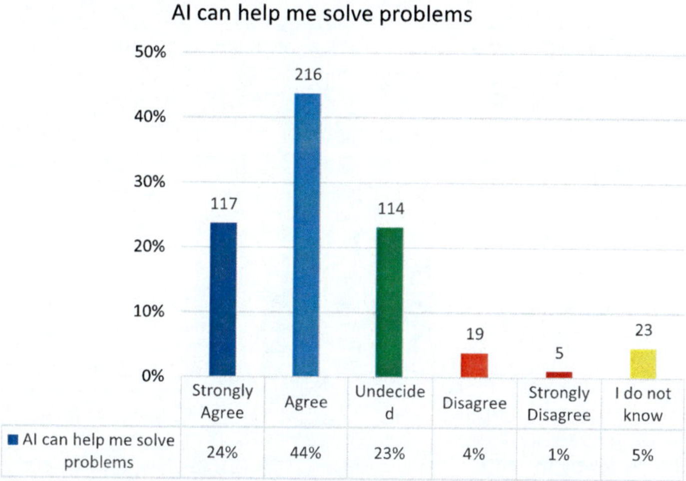

Fig. 11.15 At work, AI can help me solve problems. *Source* Author's representation of quantitative results

Q5.4 Through automation, AI can help me enhance my work performance

A total of 376 agreed that automation will enhance their work performance. There were 78 and 21 who were undecided or did not know whether the statement is true or false, while 19 participants did not think that AI can help them enhance their performance (Fig. 11.17).

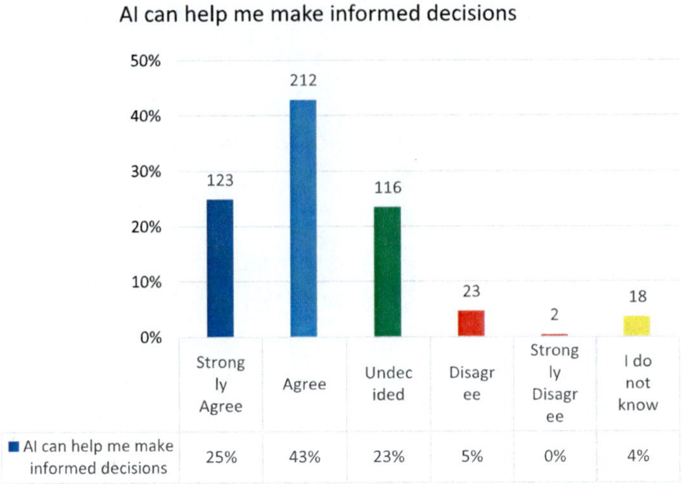

Fig. 11.16 At work, AI can help me make informed decisions. *Source* Author's representation of quantitative results

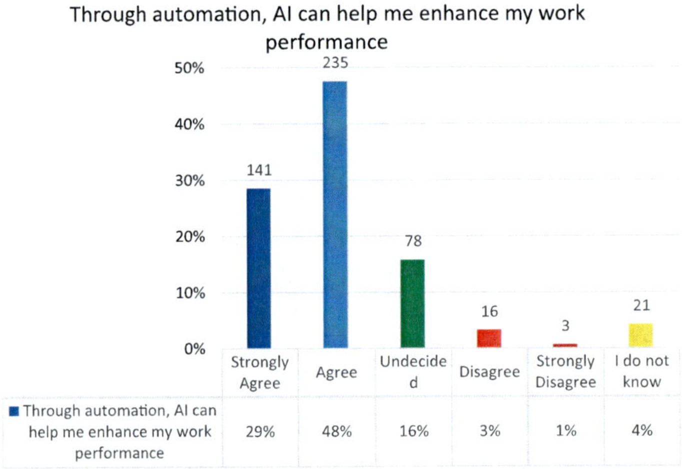

Fig. 11.17 At work, through automation AI can help me enhance my work performance. *Source* Author's representation of quantitative results

Q5.5 AI is better than humans in analysing information

The majority of respondents were undecided about whether AI can analyse information better than humans. On the other hand, a total of 171 employees agreed with the statement, while 90 disagreed and 29 strongly disagreed that AI is better at analysing information. Another 22 participants did not know how true the statement is (Fig. 11.18).

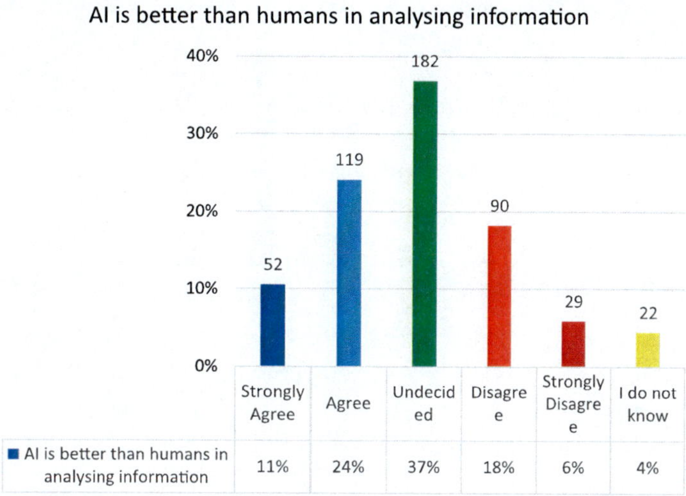

Fig. 11.18 At work, AI is better than humans in analysing information. *Source* Author's representation of quantitative results

Q5.6 I will find it difficult to learn how to use AI systems

There were 187 and 99 respondents respectively who disagreed and strongly disagreed to the statement. Another 121 were undecided, while 34 did not know whether they will find it difficult to learn how to use AI systems. Finally, 53 employees agreed that it will be difficult for them to learn how to use AI systems (Fig. 11.19).

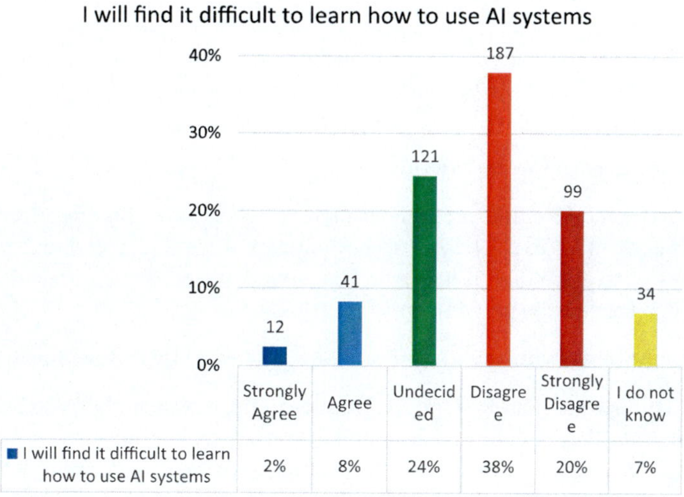

Fig. 11.19 I will find it difficult to learn to use AI systems at work. *Source* Author's representation of quantitative results

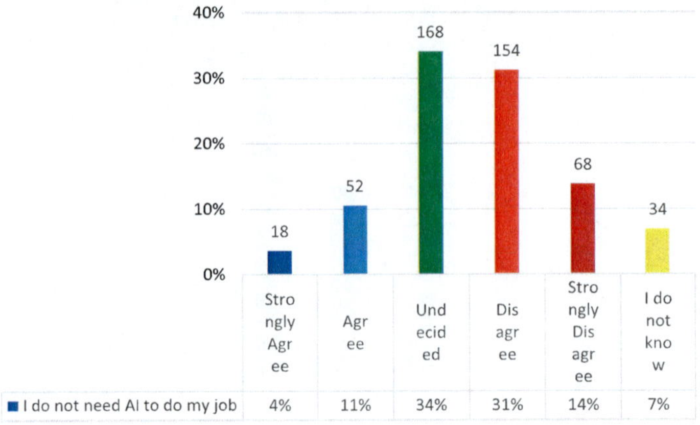

Fig. 11.20 I do not need AI to do my job. *Source* Author's representation of quantitative results

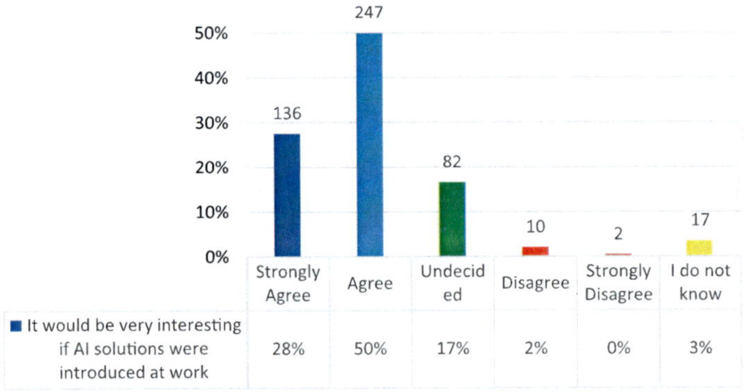

Fig. 11.21 It would be very interesting if AI solutions were introduced at work. *Source* Author's representation of quantitative results

Q5.7 I do not need AI to do my job

There were 154 and 68 employees who disagreed and strongly disagreed respectively with the statement of Fig. 11.17; 168 participants were undecided and 34 did not know if they need AI in their role, while 70 respondents agreed that they do not need AI to carry out their job (Fig. 11.20).

Q5.8 It would be very interesting if AI solutions were introduced at work

Figure 11.21 illustrates that 136 and 247 public employees strongly agreed and agreed respectively that they would find it interesting if AI technology was introduced as part of their role. On the other hand, 82 respondents were undecided, a further 17 did not know and 12 others disagreed with the statement.

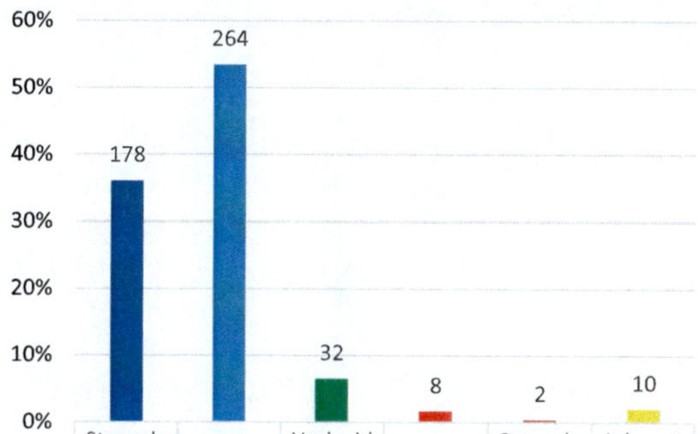

	Strongly Agree	Agree	Undecided	Disagree	Strongly Disagree	I do not know
I would like to learn more about AI uses	36%	53%	6%	2%	0%	2%

Fig. 11.22 I would like to learn more about AI uses. *Source* Author's representation of quantitative results

Q5.9 I would like to learn more about AI uses

In reply to question 5.9, a total of 442 participants agreed on the interest to learn more about the uses of AI while 10 disagreed with the statement. Another 32 public employees were undecided and a further 10 others did not know whether they would like to learn more about AI (Fig. 11.22).

Q5.10 I would like to learn how AI can help me perform better at work

Almost all of the respondents agreed in wanting to learn how AI can help them perform better at work with the exception of 30 employees who were undecided, 8 who disagreed and 8 who did not know whether they would like to learn how AI can help them perform better (Fig. 11.23).

Q6. AI Readiness among the Public Administration Workforce

Question 6 presented several statements to investigate the transformational and technical readiness as well as the organisational and environmental preparedness in relation to AI adoption. The validity of the statements was portrayed on a Likert scale.

Q6.1 Transformational/Strategic readiness

Q6.1.1 When compared to other technologies, AI is the one that will help me improve my work

Figure 11.24 illustrates that 46 public employees strongly agreed and 184 agreed; however, 189 respondents were undecided and 42 did not know whether AI is what they need to improve their work. On the other hand, 34 employees disagreed that AI is the tool they need.

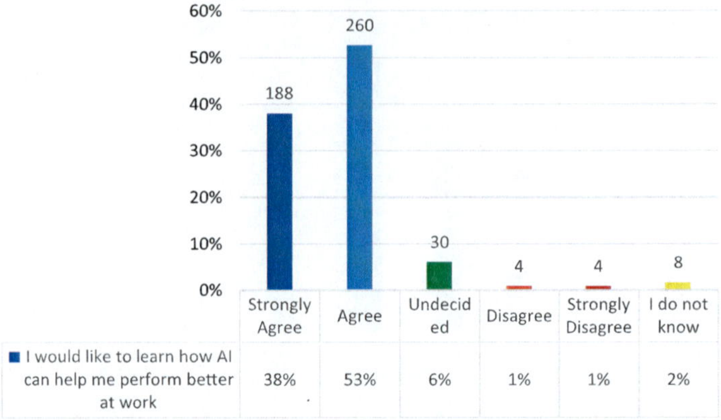

Fig. 11.23 I would like to learn how AI can help me perform better at work. *Source* Author's representation of quantitative results

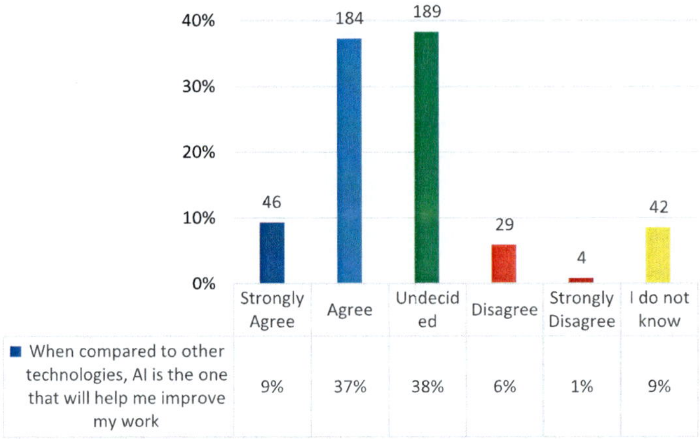

Fig. 11.24 When compared to other technologies, AI is the one that will help me improve my work. *Source* Author's representation of quantitative results

Q6.1.2 AI can facilitate collaboration with other departments and government agencies

In relation to AI facilitating collaboration, a total of 375 employees agreed that this would certainly be the case while 85 respondents were undecided and 39 others ticked the "I do not know" option. Five participants disagreed that AI could foster collaboration with other departments and government agencies (Fig. 11.25).

Q6.1.3 I am ready to use AI technologies in my day-to-day operations

Out of the total respondents, 110 strongly agreed and a further 260 were ready to use AI solutions in their daily tasks. There were 93 respondents who were undecided

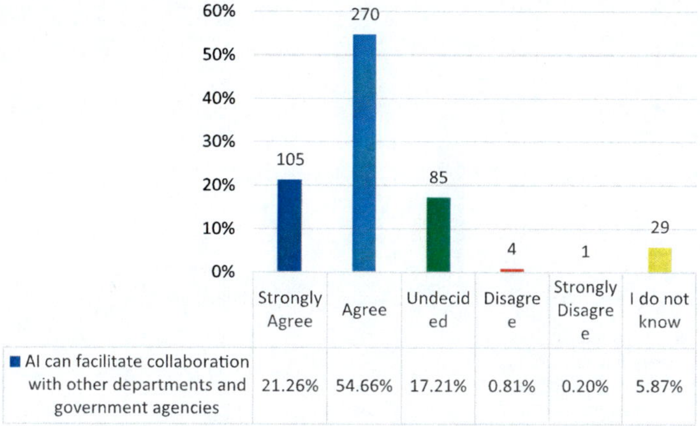

Fig. 11.25 AI can facilitate collaboration with other departments and government agencies. *Source* Author's representation of quantitative results

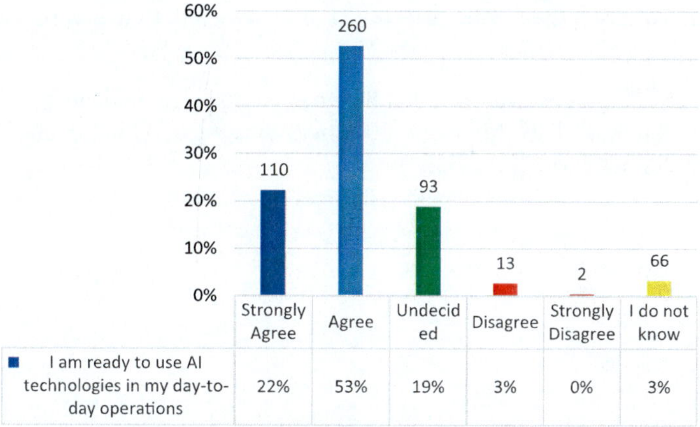

Fig. 11.26 I am ready to use AI technologies in my day-to-day operations. *Source* Author's representation of quantitative results

and 16 others who did not know. The remaining 15 participants were not ready to use AI as part of their day-to-day operations (Fig. 11.26).

Q6.1.4 I will support the government's strategy towards change in implementing AI

Out of 494 participants, 424 would support the government in implementing the change needed towards AI adoption, while 5 disagreed with the statement, 57 were undecided and another 8 selected the "I do not know" option (Fig. 11.27).

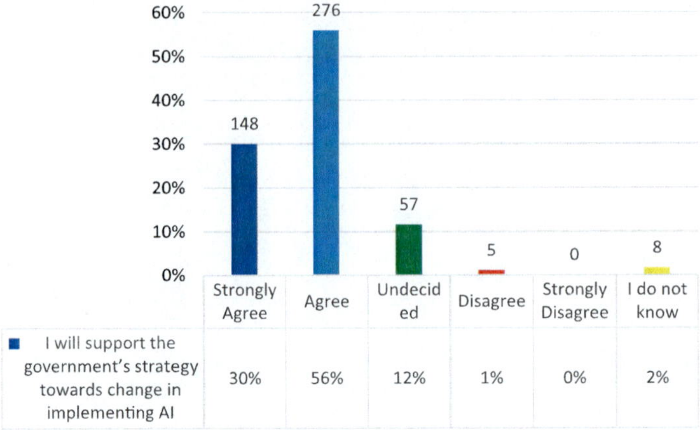

Fig. 11.27 I will support the government's strategy towards change in implementing AI. *Source* Author's representation of quantitative results

Q6.1.5 I need assistance in identifying those processes that can be enabled by AI

A total of 380 respondents agreed that they need assistance in identifying processes that can be facilitated by AI, while 30 others disagreed, 71 were undecided and another 13 did not know (Fig. 11.28).

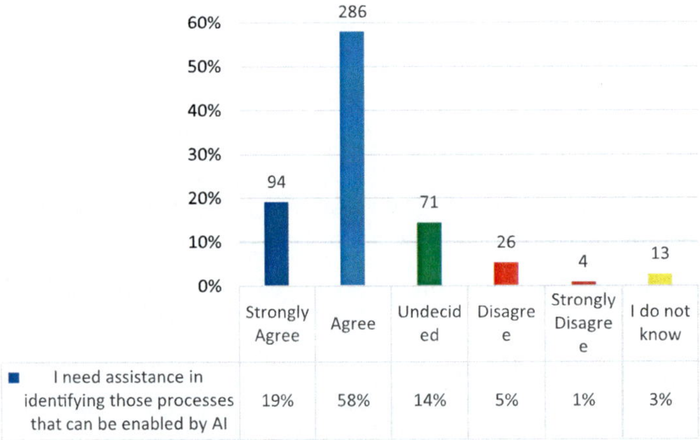

Fig. 11.28 I need assistance in identifying those processes that can be enabled by AI. *Source* Author's representation of quantitative results

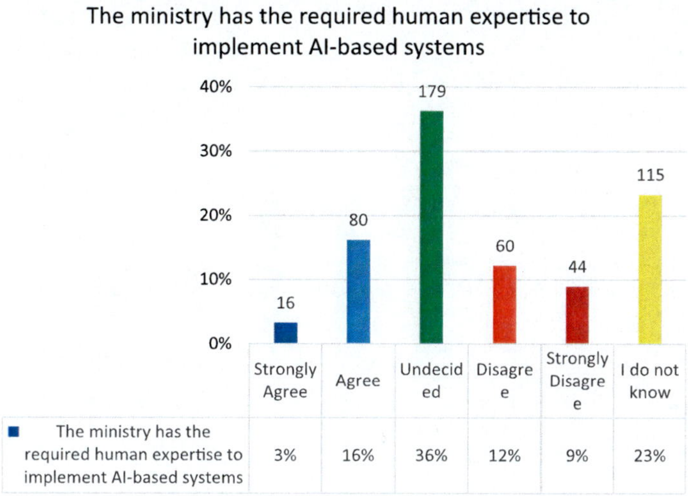

Fig. 11.29 The ministry has the required human expertise to implement AI based systems. *Source* Author's representation of quantitative results

Q6.2 Technical readiness

Q6.2.1 The ministry has the required human expertise to implement AI-based systems

Figure 11.29 presents 96 participants agreeing to the statement, while the majority of 179 were undecided. There were 60 employees who disagreed and 44 who strongly disagreed, while 115 others did not know whether the ministry has the required human expertise to implement AI.

Q6.2.2 The ministry has the required technical knowledge to operate AI based systems

Out of the total, 177 participants were undecided and 110 others did not know whether the ministry has the required technical knowledge to implement AI-based systems. On the other hand, a total of 104 respondents agreed and 103 others disagreed (Fig. 11.30).

Q6.2.3 The infrastructure to run AI systems is already in place

The majority of respondents, 187, were undecided as to whether the infrastructure to run AI systems is in place and 132 chose "I do not know". However, 56 participants agreed to the statement while 119 disagreed that the infrastructure is already in place (Fig. 11.31).

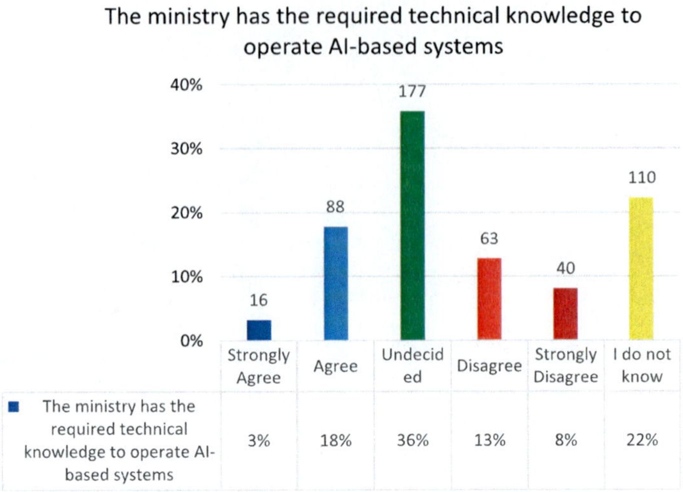

Fig. 11.30 The ministry has the required technical knowledge to operate AI based systems. *Source* Author's representation of quantitative results

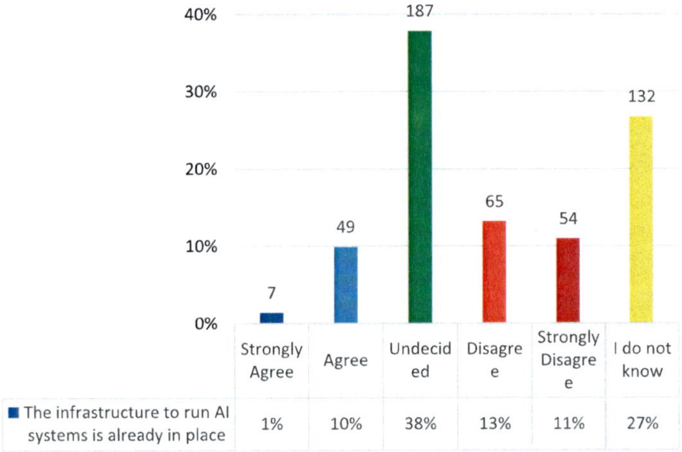

Fig. 11.31 The infrastructure to run AI systems is already in place. *Source* Author's representation of quantitative results

Q6.3 Organisational readiness

Q6.3.1 Communication is essential in getting the public workforce ready for AI adoption

As per Fig. 11.32, 463 participants agreed that communication is essential; however, 3 others disagreed with the statement. There were 19 public employees who were

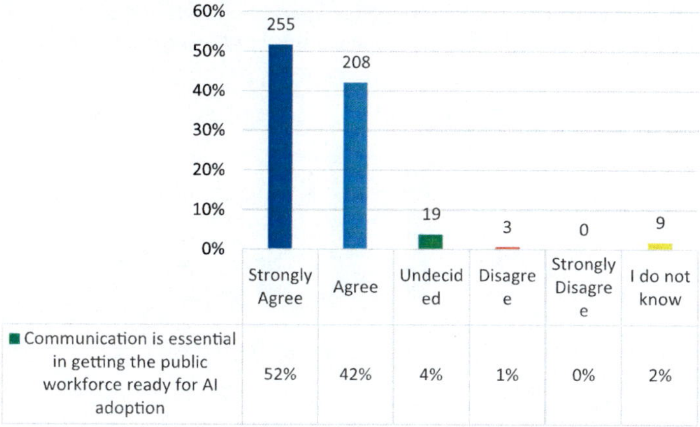

Fig. 11.32 Communication is essential in getting the public workforce ready for AI adoption. *Source* Author's representation of quantitative results

undecided and 9 did not know if communication will help get the public workforce ready for AI adoption.

Q6.3.2 Information and knowledge sharing on AI reduces uncertainty regarding the adoption of AI as part of my role

The majority, 407 participants, agreed or strongly agreed that information and knowledge sharing can reduce uncertainty on AI. There were 7 who disagreed with the statement, while 54 and 26 were identified as undecided or "do not know" respectively (Fig. 11.33).

Q6.3.3 I am informed on how the ministry will implement AI

Regarding this statement, 170 and 124 participants disagreed and strongly disagreed respectively. Another 90 were undecided, and 74 others did not know. The final 36 respondents were informed on how the ministry will implement AI (Fig. 11.34).

Q6.3.4 I understand that a change in work processes is required

The majority of the participants, 389 in total, understood that in order to adopt AI, a change in work processes would be required. Another 62 were undecided, and 25 did not know. A total of 18 employees disagreed with the statement (Fig. 11.35).

Q6.3.5 I will trust the change AI will bring along if I am informed on the impact AI will have on my job

Figure 11.36 presents a total of 388 participants confirming that they would trust the change AI will bring about if they were informed on how this would impact their work-related duties. Another 69 were undecided and 23 did not know. Fourteen respondents disagreed with the statement.

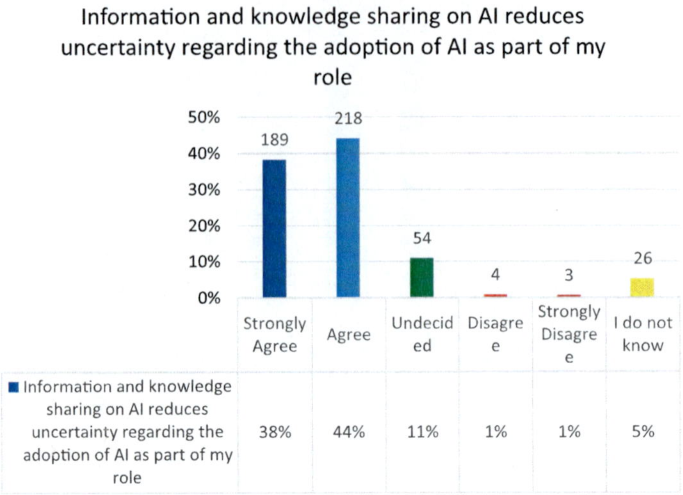

Fig. 11.33 Information and knowledge sharing on AI reduces uncertainty. *Source* Author's representation of quantitative results

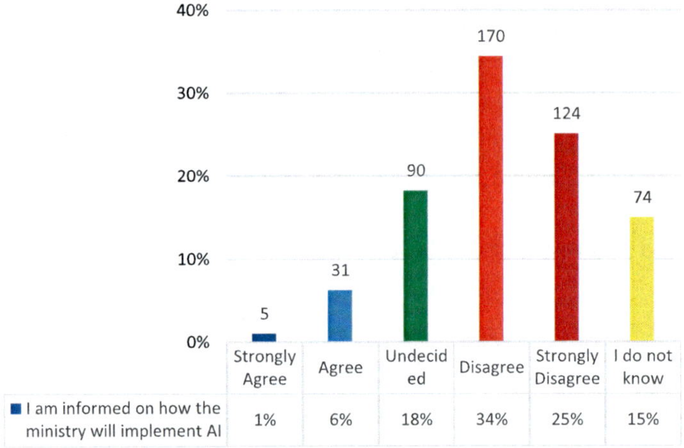

Fig. 11.34 I am informed on how the ministry will implement AI. *Source* Author's representation of quantitative results

Q6.4 Environmental readiness

Q6.4.1 There are personnel within the ministry who can help implement AI

Of the total, 147 participants agreed, while 167 were undecided as to whether the ministry has the personnel required to implement AI. Another 30 disagreed and 20 strongly disagreed, while 130 participants said that they did not know (Fig. 11.37).

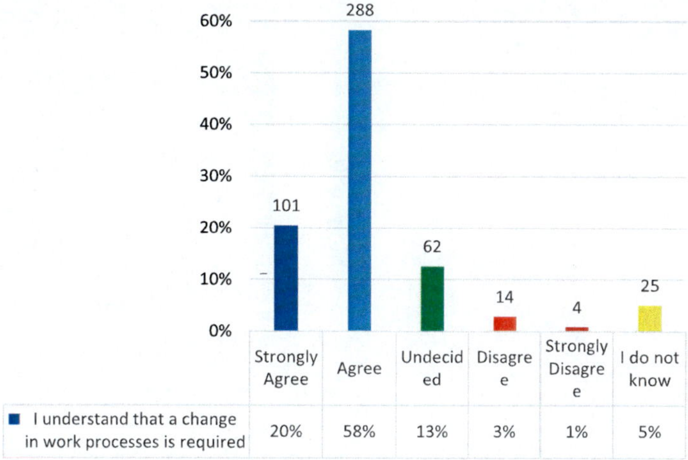

Fig. 11.35 I understand that a change in work processes is required. *Source* Author's representation of quantitative results

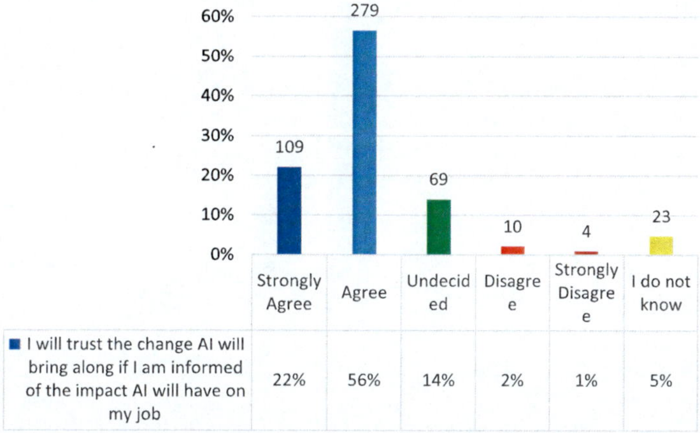

Fig. 11.36 I will trust the change AI will bring along if I am informed of the impact AI will have on my job. *Source* Author's representation of quantitative results

Q6.4.2 Third party experts can help implement AI

A total of 362 participants agreed or strongly agreed that third-party experts can help implement AI, while 10 others disagreed with the statement. There were 74 respondents who were undecided and 48 employees who did not know if the statement is valid (Fig. 11.38).

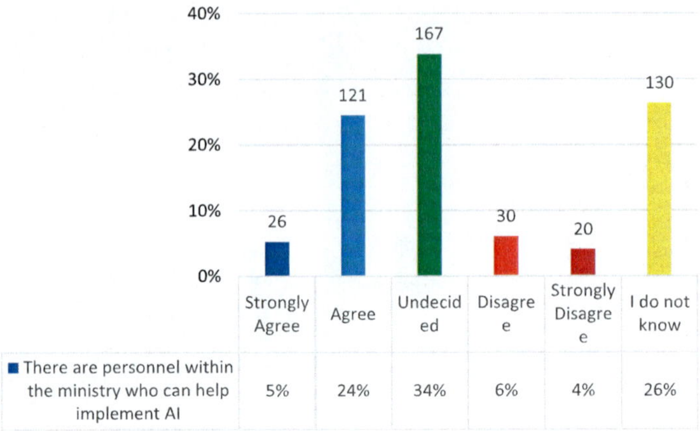

Fig. 11.37 There are personnel within the ministry who can help implement AI. *Source* Author's representation of quantitative results

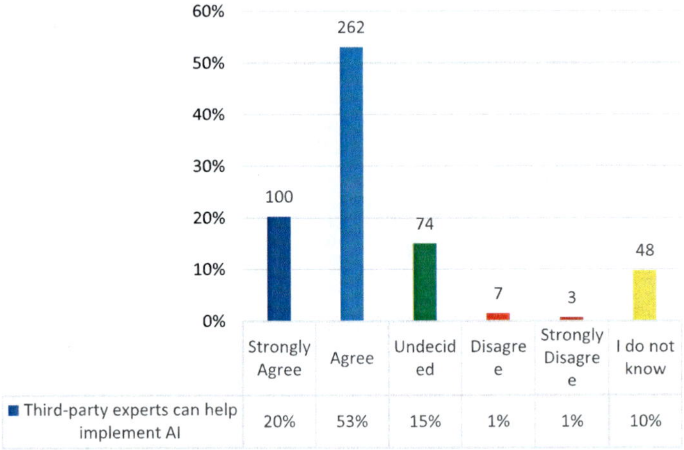

Fig. 11.38 Third-party experts can help implement AI. *Source* Author's representation of quantitative results

Q6.4.3 AI will help minimise the use of paper

A total of 403 participants agreed that AI will minimise paperwork processes. Ten respondents disagreed with this statement, while 51 and 30 were undecided or did not know respectively (Fig. 11.39).

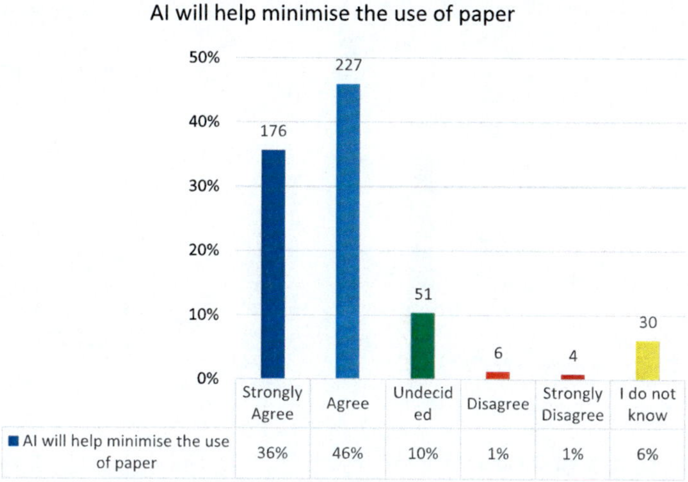

Fig. 11.39 AI will minimise the use of paper. *Source* Author's representation of quantitative results

Q6.4.4 I think the public will benefit from introducing AI systems within the public service

A major 143 and 222 respondents strongly agreed and agreed with the statement respectively. Another 85 were undecided, while 10 disagreed and a further 31 participants did not know how true the statement is (Fig. 11.40).

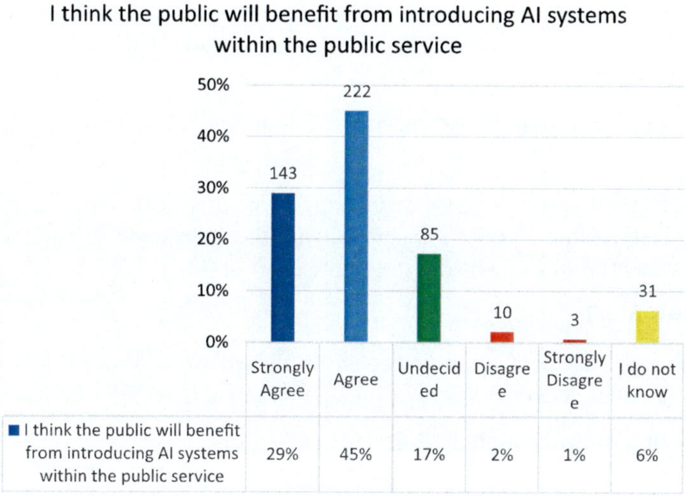

Fig. 11.40 The public will benefit from introducing AI systems within the public service. *Source* Author's representation of quantitative results

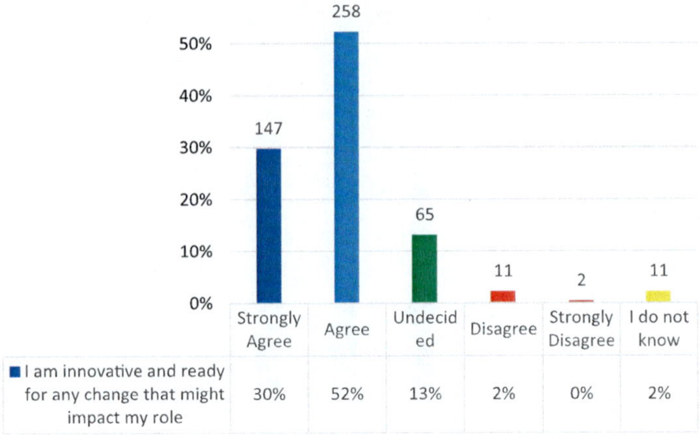

Fig. 11.41 I am innovative and ready for any change that might impact my role. *Source* Author's representation of quantitative results

Q6.4.5 I am innovative and ready for any change that might impact my role

The majority of respondents were ready for the change as 147 and 258 strongly agreed and agreed respectively. There were 65 participants who were undecided, a total of 13 who disagreed, and 11 did not know if they were ready for a change that might impact their role (Fig. 11.41).

Q6.4.6 I am aware that other ministries are using AI

Regarding this statement, the majority, 162 participants, did not know whether other ministries are using AI systems while 146 others were undecided. A total of 76 were not aware, while a total of 110 respondents were aware of the use of AI in other ministries (Fig. 11.42).

Q6.4.7 I am aware that other countries are using AI in their public service delivery

A total of 268 participants were aware that other countries are using AI in their public service delivery, while 25 others disagreed with the statement. There 98 undecided employees and 103 others who did not know (Fig. 11.43).

Q7. Benefits of AI

Question 7 explored how AI is perceived by the public administration workforce through a series of statements, whose understanding is reflected on a Likert scale.

Q7.1 AI will increase collaboration with colleagues, external stakeholders and the public

The majority of participants, that is, 343 employees, agreed that AI will foster collaboration, while 16 disagreed in this regard. There were 91 who were undecided and

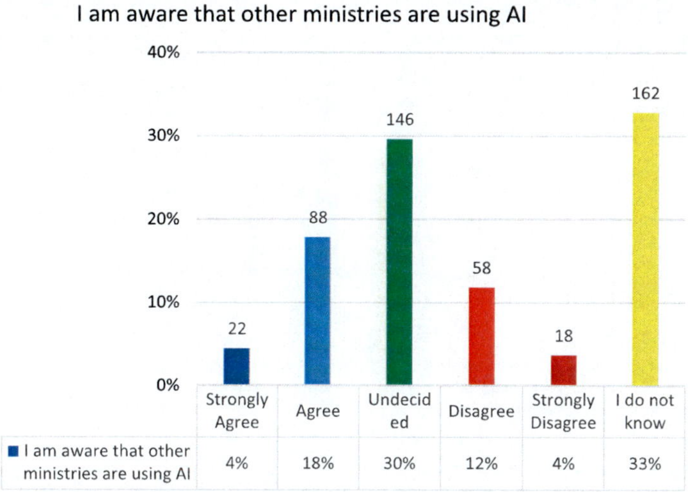

Fig. 11.42 I am aware that other ministries are using AI solutions. *Source* Author's representation of quantitative results

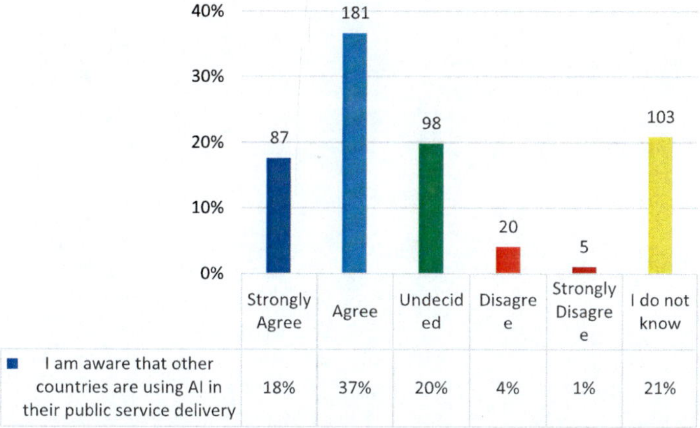

Fig. 11.43 I am aware that other countries are using AI in their public service delivery. *Source* Author's representation of quantitative results

44 who did not know if AI will increase collaboration internally and with external stakeholders (Fig. 11.44).

Q7.2 AI provides data for informed decision making

It was a collective agreement, with a total of 390 respondents approving that AI provide data for making informed decisions. There were 57 participants who were undecided and 40 others who did not know. Finally, 7 disagreed with the statement (Fig. 11.45).

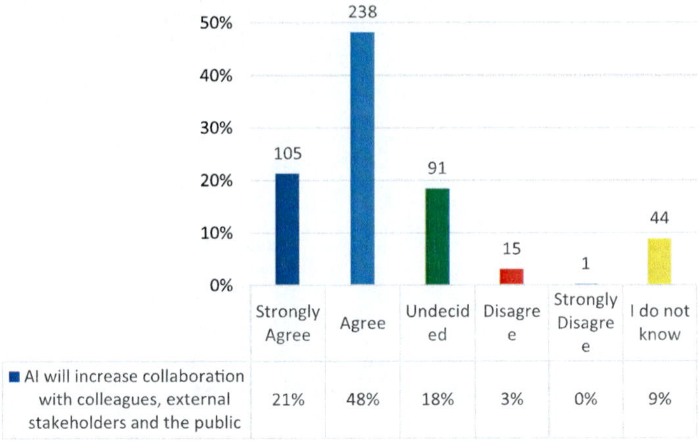

Fig. 11.44 AI will increase collaboration with colleagues, external stakeholders and the public. *Source* Author's representation of quantitative results

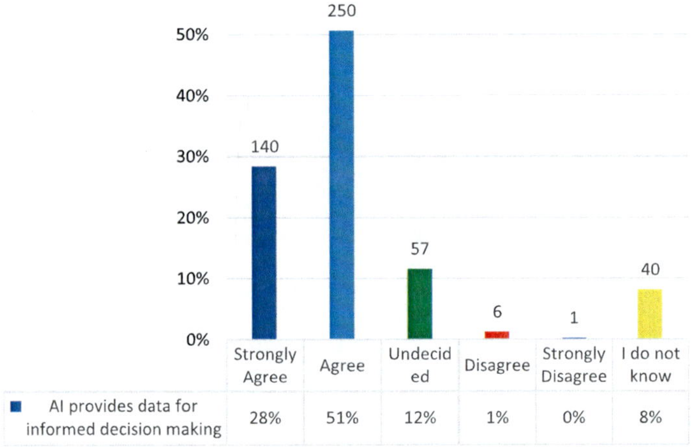

Fig. 11.45 AI provides data for informed decision making. *Source* Author's representation of quantitative results

Q7.3 AI is more reliable and consistent

Regarding this statement, 87 respondents strongly agreed and 221 others agreed. There were 126 participants who were undecided, a total of 21 who disagreed and 39 who did not know if AI is more reliable and consistent (Fig. 11.46).

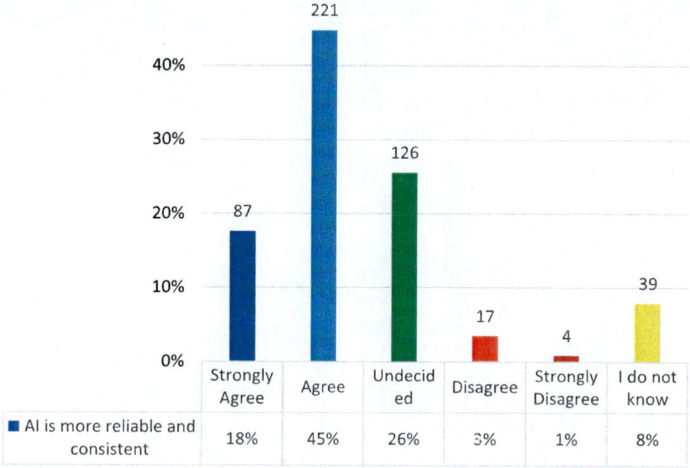

Fig. 11.46 AI is more reliable and consistent. *Source* Author's representation of quantitative results

Q7.4 AI will eliminate repetitive tasks

A total of 385 participants agreed that AI will eliminate repetitive tasks while 10 disagreed with the statement. Another 65 were undecided, while 34 respondents selected the "I do not know" option (Fig. 11.47).

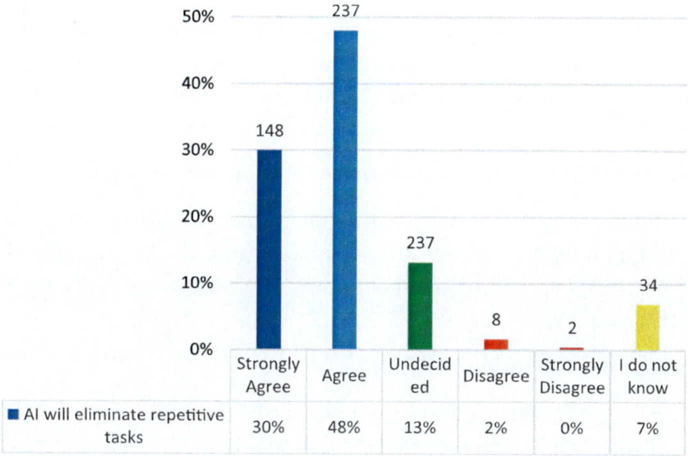

Fig. 11.47 AI will eliminate repetitive tasks. *Source* Author's representation of quantitative results

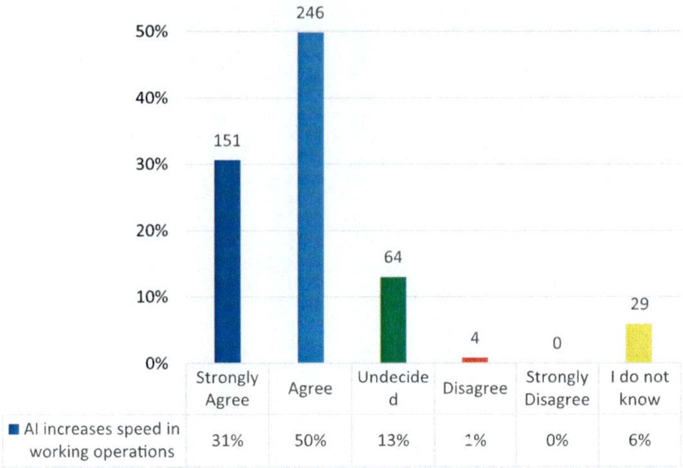

Fig. 11.48 AI increases speed in working operations. *Source* Author's representation of quantitative results

Q7.5 AI increases speed in working operations

Figure 11.48 shows that 397 respondents agreed that AI will increase the speed in working operations. While 64 participants were undecided and 29 did not know, 4 others disagreed with the statement.

Q7.6 AI will enable multi-tasking and eases the workload for public workforce

For question 7.6, there also seems to be a general agreement to the statement, with 137 and 232 respondents strongly agreeing and agreeing respectively. Another 79 were undecided, 35 did not know and 11 of the participants disagreed that AI would facilitate multi-tasking and ease the workload of the public workforce (Fig. 11.49).

Q7.7 AI will help better understand patterns in the various public service sectors

In relation to understanding patters within the public service, it was agreed that AI can assist in this regard by a total of 384 participants. There were 69 respondents who were undecided and 33 who did not know. Eight employees disagreed with the statement (Fig. 11.50).

Q7.8 AI will assist the public 24/7

A total of 370 respondents agreed that AI will assist the public 24/7, while 10 others disagreed. There were 71 who were undecided and 43 others who identified with the "I do not know" option (Fig. 11.51).

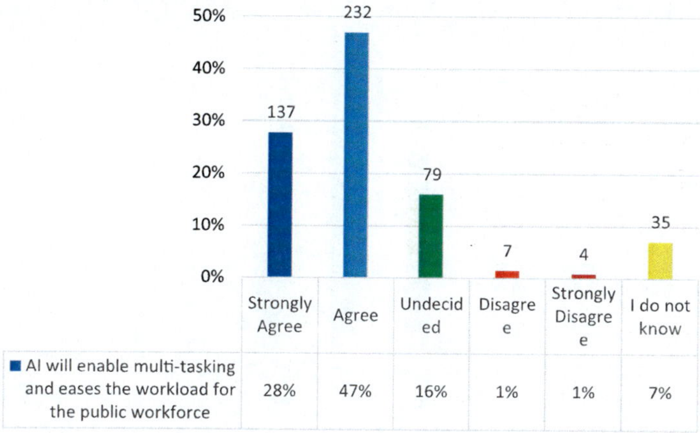

Fig. 11.49 AI enables multi-tasking and eases the workload for the public workforce. *Source* Author's representation of quantitative results

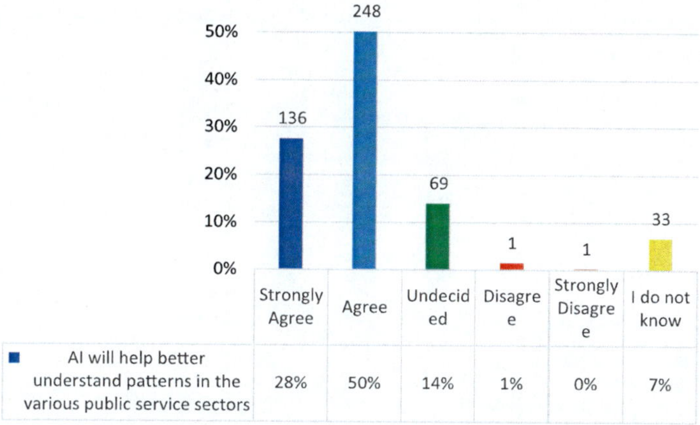

Fig. 11.50 AI will help better understand patterns in the various public service sectors. *Source* Author's representation of quantitative results

Q7.9 AI will facilitate faster communication and response with the public

A total of 391 participants agreed that AI will facilitate faster communication and response with the public, while 6 disagreed with this statement. Another 60 respondents were undecided, and 37 did not know (Fig. 11.52).

Q8. Potential risk with the introduction of AI that will hinder AI progress

Question 8 investigated the risks and possible fears that the public administration might be facing with the introduction of AI solutions. The selected statements were

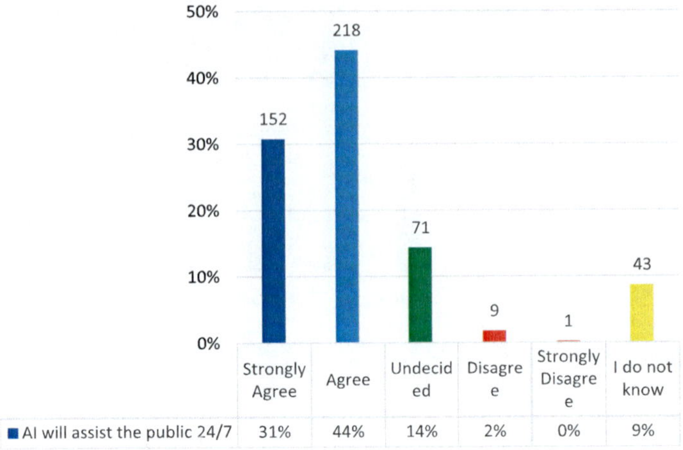

Fig. 11.51 AI will assist the public 24/7. *Source* Author's representation of quantitative results

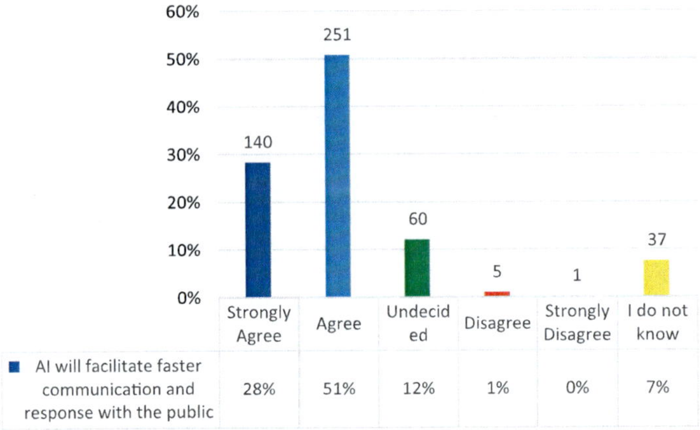

Fig. 11.52 AI will facilitate faster communication and response with the public. *Source* Author's representation of quantitative results

backed with literature and research, and the perceived impact was presented on a Likert scale.

Q8.1 I am employed with the government; hence my job is secure and I do not feel threatened by AI technologies

Although a total of 258 participant agreed with the statement, 78 and 15 respondents disagreed and strongly disagreed respectively and feel threatened by AI technologies replacing their roles. Another 112 were undecided and 31 did not know (Fig. 11.53).

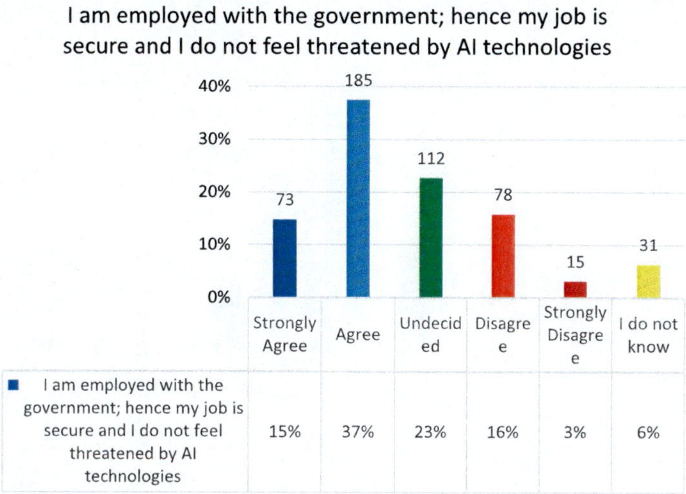

Fig. 11.53 I am employed with the government; hence my job is secured. *Source* Author's representation of quantitative results

Q8.2 I worry my job can be replaced by AI systems

Figure 11.54 portrays a total of 314 participants disagreeing with the statement, while 65 others agreed that they worry their job can be replaced by AI. Another 87 were undecided and 28 did not know.

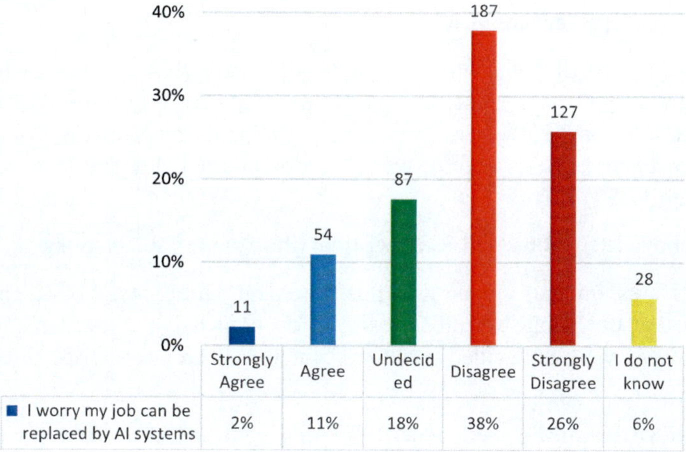

Fig. 11.54 I worry my job can be replaced by AI systems. *Source* Author's representation of quantitative results

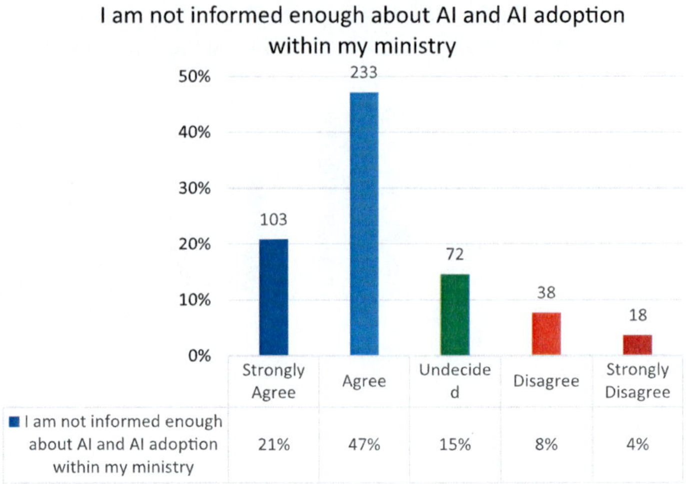

Fig. 11.55 I am not informed enough about AI adoption within my ministry. *Source* Author's representation of quantitative results

Q8.3 I am not informed enough about AI and AI adoption within my ministry

With 306 participants in agreement, there seems to be a general consensus that the participants were not informed enough about AI and AI adoption within their ministry, with 72 respondents undecided, and 38 and 18 others disagreeing and strongly disagreeing respectively (Fig. 11.55).

Q8.4 I believe the way data is currently collected by our ministry could be a barrier to the application of AI

There were 183 participants who were undecided about this statement, while a total of 149 agreed that the current data collection methods might hinder the application of AI solutions. However, 55 and 12 participants respectively disagreed and strongly disagreed in this regard, while 95 others opted for the "I do not know" option (Fig. 11.56).

Q8.5 I believe the public will feel sceptical about using AI systems

In total, 270 respondents agreed with this statement, while a total of 50 participants disagreed that the public would feel sceptical about using AI systems. However, 122 employees were undecided about this statement and a further 52 did not know (Fig. 11.57).

Q8.6 I believe the public needs more information about AI to trust new processes

There were 226 participants who strongly agreed and a further 219 who agreed that the public needs more information to trust AI processes. There were 28 respondents who were undecided, 8 who disagreed with the statement, and another 13 who selected the "I do not know "option (Fig. 11.58).

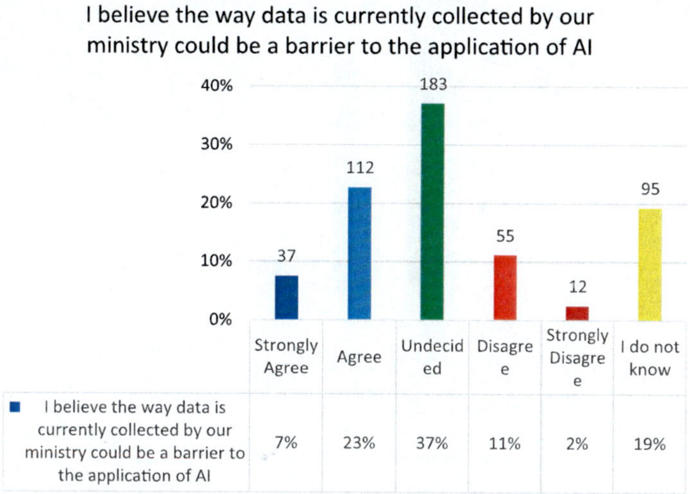

Fig. 11.56 I believe the way data is collected could be a barrier to the application of AI. *Source* Author's representation of quantitative results

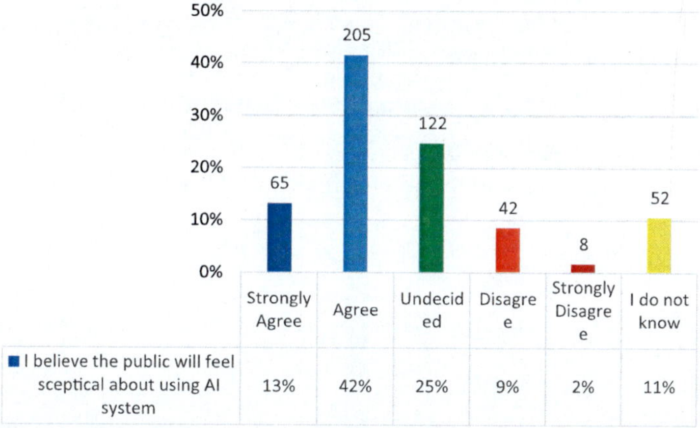

Fig. 11.57 I believe the public will feel sceptical about using AI systems. *Source* Author's representation of quantitative results

Q9. Possible changes that will take place with the introduction of AI

Question 9 analyses how accustomed participants were to the changes that will be introduced once AI is adopted. The statements were backed by research and their response was gauged on a Likert scale.

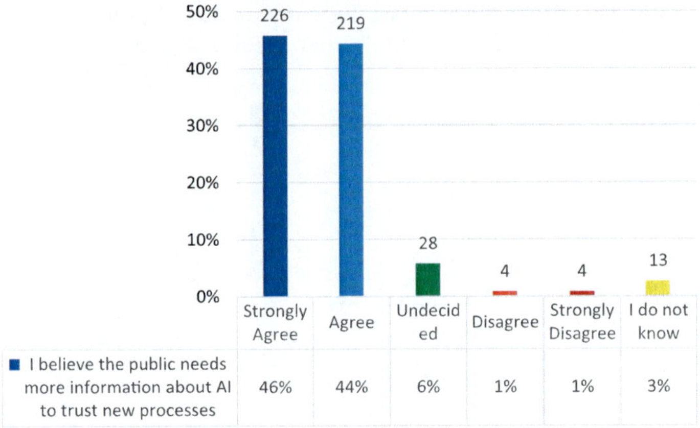

Fig. 11.58 I believe the public needs more information about AI to trust new processes. *Source* Author's representation of quantitative results

Q9.1 AI will change the current work processes

A total of 402 participants agreed that AI will transform the current work processes while 12 disagreed with the statement, 56 respondents were undecided and 24 others did not know (Fig. 11.59).

Q9.2 New jobs will be created within the Government

While 81 and 198 respondents respectively strongly agreed and agreed that there will be new job creations within the government, 133 others were undecided and

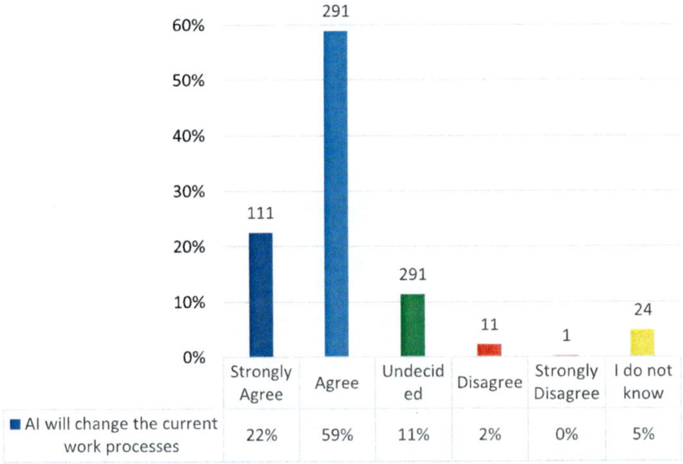

Fig. 11.59 AI will change the current work processes. *Source* Author's representation of quantitative results

34 disagreed altogether. There were also 48 participants who selected the "I do not know" option (Fig. 11.60).

Q9.3 Public administrators will need to be retrained to upskill their current role

The majority of the participants, that is, 418 in total, agreed that public administrators will need to be retrained, while 8 others disagreed with this statement, 48 were undecided and 20 did not know whether that would be necessary (Fig. 11.61).

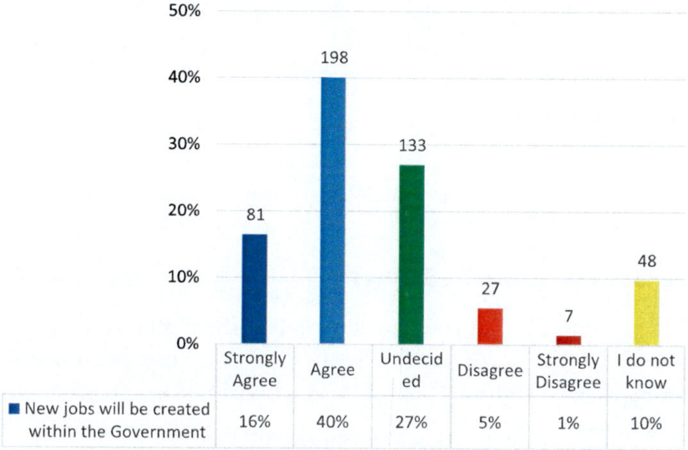

Fig. 11.60 New jobs will be created within the government. *Source* Author's representation of quantitative results

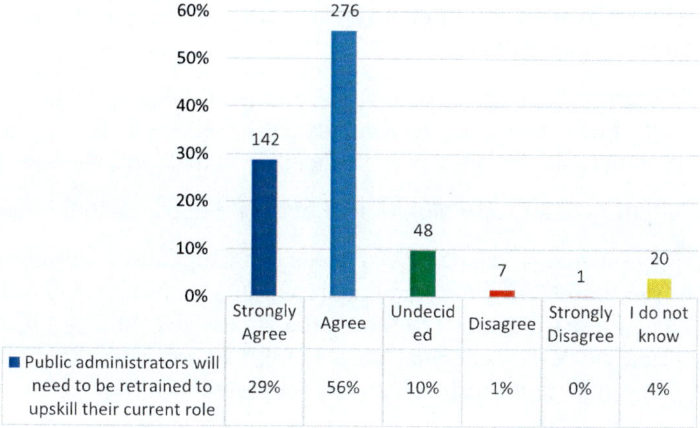

Fig. 11.61 Public administrators will need to be retrained to upskill their current role. *Source* Author's representation of quantitative results

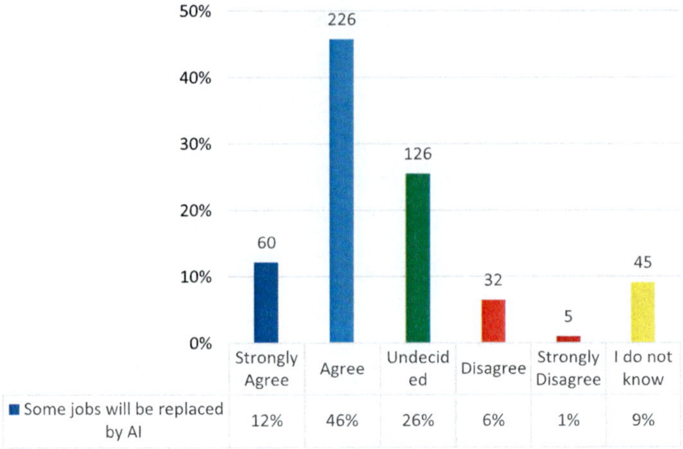

Fig. 11.62 Some jobs will be replaced by AI. *Source* Author's representation of quantitative results

Q9.4 Some jobs will be replaced by AI

A total of 286 respondents agreed that some jobs will be replaced by AI, while 126 others were undecided. There were 37 participants who disagreed and 45 who did not know whether jobs will be replaced by AI (Fig. 11.62).

Q9.5 AI will facilitate the roles of the public workforce

There seemed to be a general agreement among 348 participants that AI will facilitate the roles of the public workforce. While 97 respondents were undecided, 12 disagreed and 37 ticked the "I do not know" option (Fig. 11.63).

Q9.6 AI will address more efficiently the needs of the public workforce in delivering quality public service

A total of 357 respondents agreed that AI will efficiently address the needs of public administration with the aim to deliver a quality public service. In this regard, 97 were undecided, 31 others did not know and a further 9 participants disagreed (Fig. 11.64).

Q10. The adoption of AI technologies within the ministries will be successful if:

Question 10 speculated on the perceived needs that the public workforce envisages in order for successful acceptance and adoption of AI solutions within the public administration. These were drawn following the discussions with the CIOs and what they anticipated would be beneficial for a smooth AI transformation within their ministries. The replies of the respondents were reflected on a Likert scale as follows:

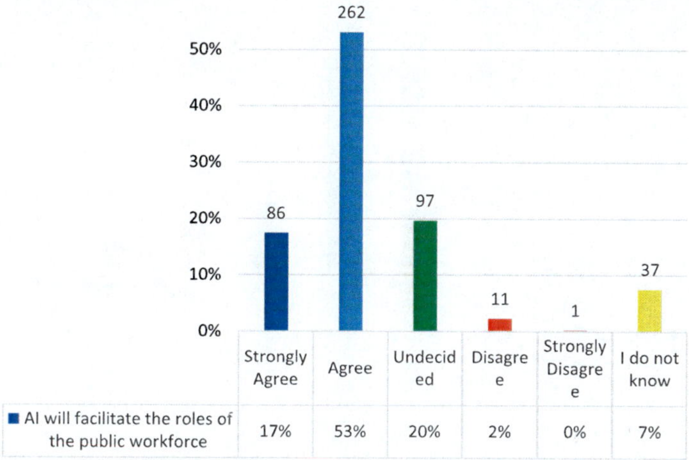

Fig. 11.63 AI will facilitate the roles of the public workforce. *Source* Author's representation of quantitative results

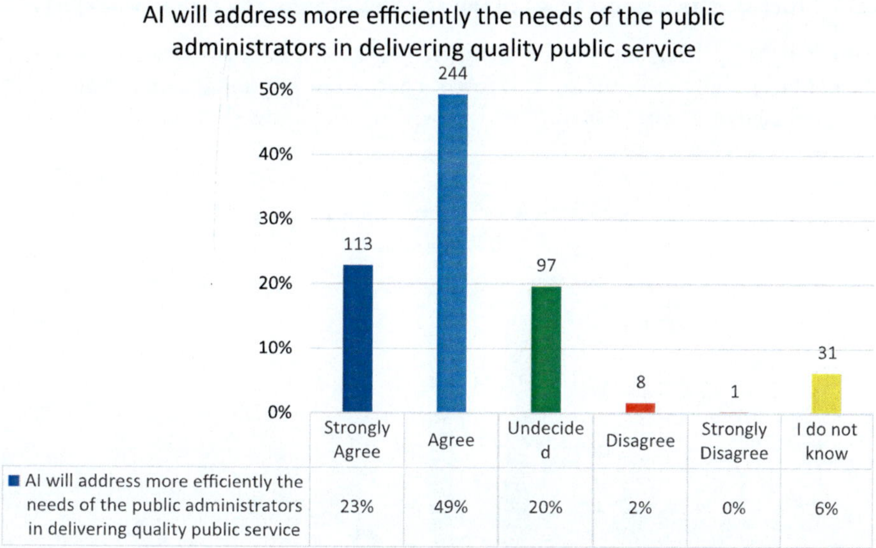

Fig. 11.64 AI will address more efficiently the needs of the public administrators. *Source* Author's representation of quantitative results

Q10.1 The public workforce is informed and trained with the right skills in the use of AI

A total of 354 of the participants agreed that for a successful AI adoption, the public workforce should be informed and trained with the right skills. Another 51 respondents were undecided about this, while a total of 70 disagreed with this statement. The other 19 participants selected the "I do not know" option (Fig. 11.65).

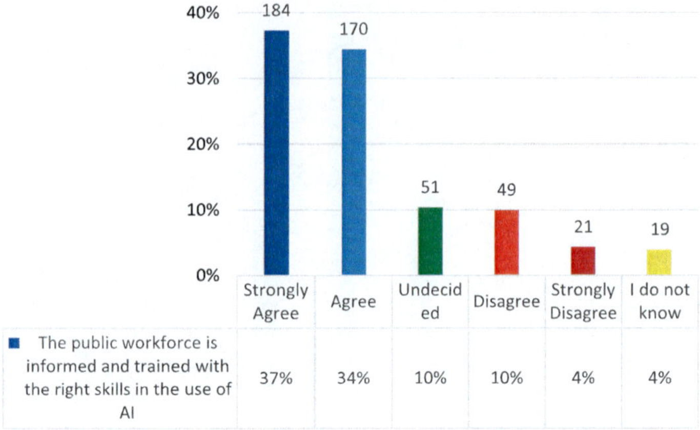

Fig. 11.65 If the public workforce is informed and trained with the right skills in the use of AI. *Source* Author's representation of quantitative results

Q10.2 Increase awareness of AI among the public and the public workforce

A total of 442 participants agreed that there needs to be increased awareness among the public as well as the public workforce. There were 31 public administrators who were undecided, 8 who disagreed and 13 who did not know (Fig. 11.66).

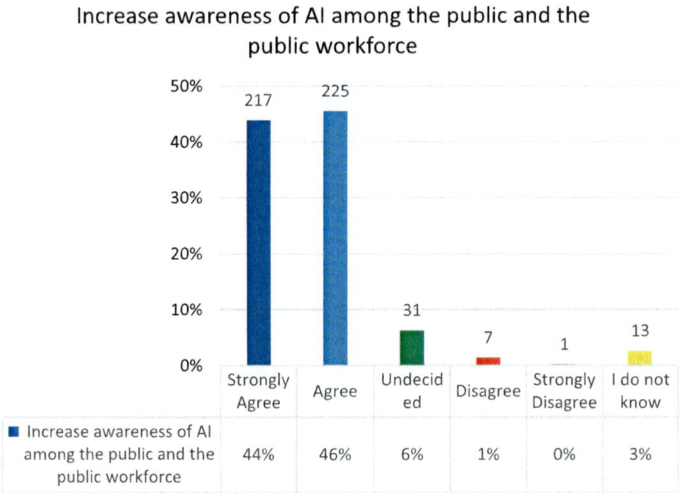

Fig. 11.66 Increase awareness of AI among the public and public workforce. *Source* Author's representation of quantitative results

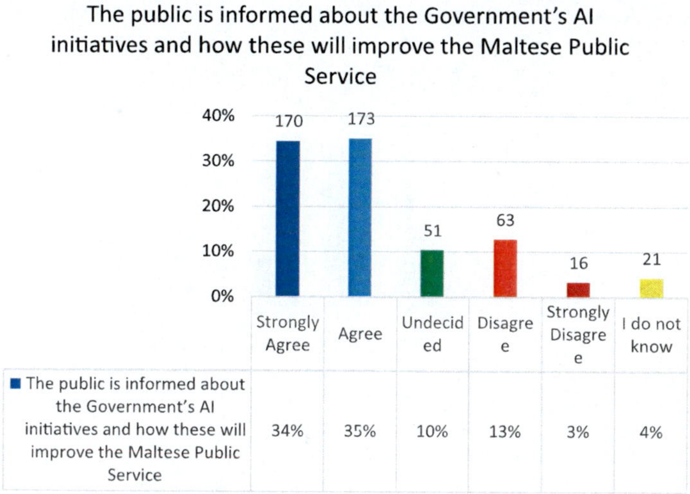

Fig. 11.67 The public is informed about the government's AI initiatives. *Source* Author's representation of quantitative results

Q10.3 The public is informed about the Government's AI initiatives and how these will improve the Maltese Public Service

A total of 343 respondents agreed that the public should be informed about the Government's AI initiatives and how these will impact the public service. Another 51 participants were undecided, while 79 disagreed with this statement. There were 21 public servants who did not know if the public should be informed (Fig. 11.67).

Q10.4 AI is introduced as part of the Maltese education system early in order to have an AI knowledgeable future workforce

Out of the 494 participants, 403 agreed that AI should be introduced in the Maltese education early in order to have an AI-knowledgeable future workforce. This will facilitate the adoption of AI at a faster pace within the public administration. However, 49 respondents were undecided, 16 disagreed with the statement and 26 others did not know (Fig. 11.68).

Q10.5 Preparation and planning for the introduction of AI is done strategically and for the long term

There was general consensus among 429 participants that preparation and planning for the long term is key for the successful implementation of AI technologies. 41 respondents were undecided while 5 others disagreed with the statement and a further 19 did not know (Fig. 11.69).

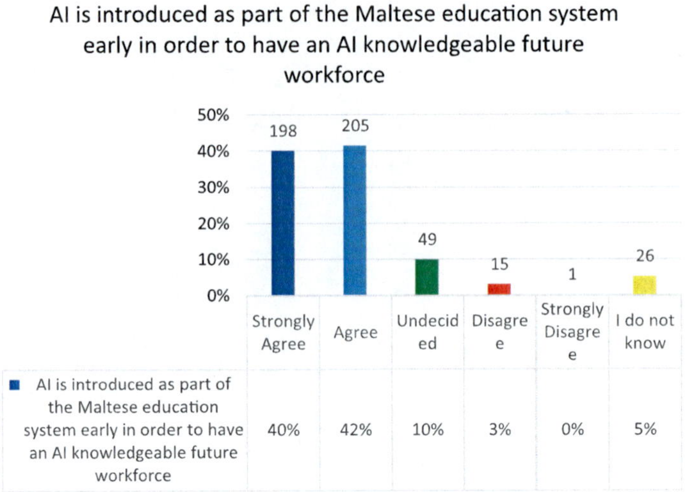

Fig. 11.68 AI is introduced as part of the Maltese education system. *Source* Author's representation of quantitative results

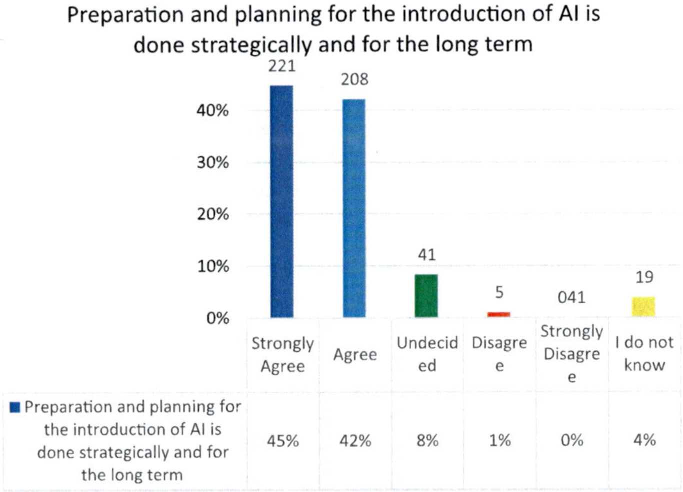

Fig. 11.69 Preparation and planning for the introduction of AI is done strategically. *Source* Author's representation of quantitative results

Q10.6 The implementation of AI systems is set against a robust legislative framework to protect data privacy while promoting transparency and accountability

A total of 402 public employees agreed with the statement that AI should be set against a robust legislative framework, while 58 respondents were undecided and 4 others disagreed. Thirty participants opted for the "I do not know" option (Fig. 11.70).

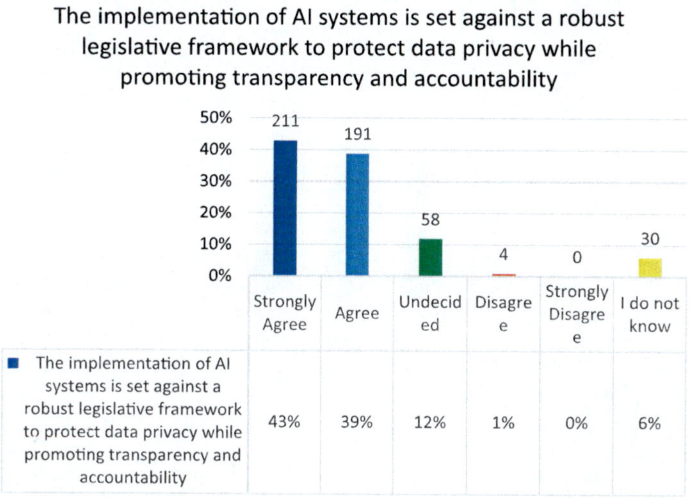

Fig. 11.70 The implementation of AI systems is set against a robust legislative framework. *Source* Author's representation of quantitative results

Q10.7 The Government starts looking into the possible future economic risks related to AI adoption that might impact the Maltese population

A total of 390 participants agreed that the government should look into the possible economical risks related to the adoption of AI and how it might impact society. Another 65 public employees were undecided and 7 others disagreed, while 33 respondents chose the "I do not know" option (Fig. 11.71).

Conclusion

The presentation of findings collated the outcomes of both the qualitative and quantitative research with the aim to investigate AI readiness within the public administration. The qualitative research data was dissected into prominent themes discussed with the CIOs and public sector personnel during the interviews. On the other hand, the survey presented the views of the varied workforce, also employed under different ministries and engaged in different scales, hence feedback was received from a wider audience, facilitating for a holistic approach to the research analysis.

11.9 Analysis

11.9.1 Introduction

This chapter presents a triangulation analysis whereby the qualitative results will be examined vis-à-vis the quantitative findings and in relation to literature on Artificial

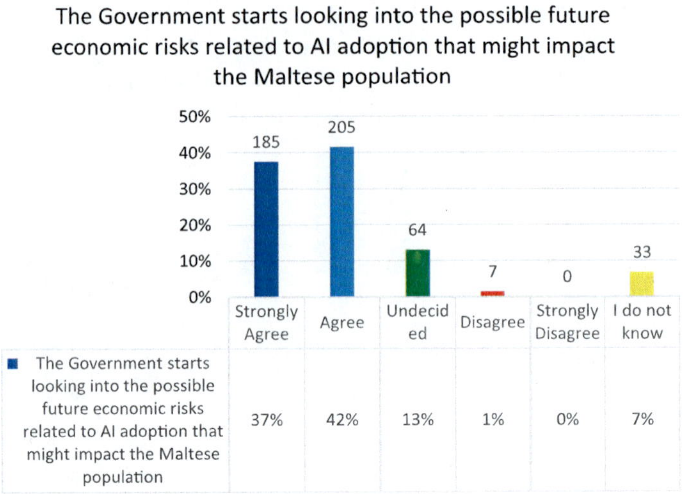

Fig. 11.71 The government starts looking into the possible future economic risks. *Source* Author's representation of quantitative results

Intelligence (AI) readiness, specifically the Malta AI Strategy [12–18]. The aim of the triangulation approach is to enhance validity, develop a more detailed perception of the research topic, and to find alternative ways of interpreting the research problem [19]. Hence, the triangulation analysis will establish complementarity, convergence or divergence, and consequently identify concepts where further investigation is required.

To validate the qualitative model, the researcher will present and discuss the interview results via a thematic approach, whereby each identified overarching theme will be further analysed in view of the survey findings. A full comprehensive list of excerpts to further corroborate the analysis is presented in Appendices E–N. Finally, a set of recommendations, based on the research findings and the Malta AI Strategy [12–18] and other relevant literature, will be proposed in the next chapter.

11.9.2 Understanding AI

This section discusses the first overarching theme, 'Understanding AI', which addresses two (2) themes; 'AI as a Tool' and the 'Current uses of AI within the Maltese government'.

11.9.2.1 Theme: AI as a Tool

The interviewees agree that there is no universal definition of AI, however, it is considered as a tool that can meet the need of the different departments within the ministries and enhance public services, through the autonomous implementation of certain human cognitive abilities.

Interviewee 14

> AI is a tool that will help and accommodate customers, clients, public administration, to enhance business services.

This is in line with the European Commission's efforts to define AI as a tool which is transforming a wide variety of sectors by automating operations that would normally require human intelligence to complete [20]. As such, AI contributes to complicated scientific and technical operations by efficiently replicating, complementing, or enhancing human intellect [21]. To this effect, 72% of the survey respondents identify AI as a tool related to intelligence, robotics and machine learning, computer programs that learn and can take decisions, smart technologies, algorithms generating information and systems that perform tasks without human input (Table 11.1). Hence, the majority of research participants understand the different contexts related to AI as a tool, which is essential in the process of AI adoption.

11.9.2.2 Theme: Current Uses of AI

Numerous public sector organizations worldwide are already effectively utilising AI for duties ranging from fraud detection to customer service [22]. The interviewees commented on the investment in cloud solutions by the Maltese government, and on how a number of ministries are now further exploiting to reap maximum benefits out of this tool. Business intelligent systems and business portals are being procured to maximise data warehousing and the provision of analytics regarding both employees and the public. A number of ministries are already using AI systems in their operations.

Interviewee 04

> the ministry has numerous systems in place targeting project management, resource allocation or resource management.

For instance, chatbots are being used to facilitate communication both internally; in seeing to staff queries and externally; in assisting the general public in their enquiries. Additionally, interviewees confirmed that a number of Internet of Things (IoT) systems are in place to inform the public and assist the authorities in drawing up an action plan on the areas being investigated.

When asked whether AI technologies were already available in their respective ministry (Fig. 11.6), 33% of the survey respondents assert that there are no AI technologies in use while 53% are not sure. However, 14% confirm that AI is being applied in their respective ministry. According to the theme classification of the replies, the current use cases of AI were mostly identified in the medical (33%) and agricultural (14%) fields (Table 11.2). Nevertheless, the Government through the Malta AI Strategy [12–18] has rolled out six (6) high-profile pilot projects across different ministries, aimed at delivering better public services and enhancing public administration operations. To this effect, further awareness is required to inform the public officers of the current adoption of AI technologies within their respective ministry and across the whole Government. Furthermore, the status quo of the aforementioned pilot projects needs to be determined, including feedback received and hindering factors that might have impacted the initiative. This will give the necessary underpinnings in view of the potential viability of a nationwide rollout.

11.9.3 Policy and Procedures

This section discusses the second overarching theme, 'Policy and Procedures', which addresses two (2) themes; the Malta AI Strategy [12–18] and the 'Ministry's Own' internal AI policies.

11.9.3.1 Themes: The Malta AI Strategy [12–18] and Ministry's Own

In general, the interviewees agree that the Malta AI Strategy [12–18] is an excellent starting point with concrete deliverables. Two particular interviewees mentioned that even though Malta's achievement about the strategy so far has been remarkable, the strategy itself lacks an implementation plan, an immediate vision and a strategic plan for the immediate future.

Interviewee 18

> The strategy is 10yrs long and the document is clear on what the deliverables will be, however, it is missing a concrete implementation plan.

The survey established that 75% of the respondents are not aware that the Government launched an AI Strategy [12–18], pronouncing the lack of information among the workforce (Fig. 11.7). The remaining 25% are aware of the strategy and commented on how AI will impact their role (Table 11.3). The greatest impact by the survey respondents is seen in terms of increased workflow efficiency (27%). Hence, even though the interviewees are aware of the AI solutions that exist or will be introduced, it seems that such information is still at the top levels, notwithstanding the informative campaign that ought to have taken place, according to the Malta AI Strategy [12–18]. Nevertheless, from the survey responses, it seems that the Government is still to implement the series of awareness events that were due between 2020 and 2021. Preparing the future workforce and controlling the emergence of AI inclusively can help maximize the benefits of new technologies while mitigating employment-related risks. Public officers can flourish in an AI-driven economy, but governments must assist them to reach their full potential [23].

Through the interviews, it transpires that only a handful of ministries have an AI deployment strategy in place.

Interviewee 04

I, as a CIO, I have to have a strategy. I drew a strategic document, which is tagged with the level of organisational chart that I wish to have, with the level of work of each level, of how it is going to affect my unit. Basically, the more systems you have, the more management and policies you have to adapt.

This appears to be a motivating factor and driving force in the implementation and application of AI technologies in their ministry. However, the majority of the ministries have no strategy in place as they prefer to await the AI technology to be available at their ministry, before drawing up a document.

Interviewee 09

Currently, there is no strategy in place. A strategy will be drawn up once the processes are being implemented.

It is also commented that usually within the public service, things are not approached holistically. However, those with an in-house strategy in place are making evidence-based decisions while pre-empting risks on how AI could impact their department.

Interviewee 06

... how things work in the public service most of the times is that, there's not like a holistic view of what is going to happen.

11.9.4 Data

This section discusses the overarching theme, 'Data', which looks into themes regarding the data availability; including the data collection processes, structures in place, data quality and dumping. Furthermore, it discusses data processes related to the legal frameworks and ethical AI practices.

11.9.4.1 Data Availability

The adoption of AI significantly exceeds the boundaries of existing data processing and analytical capability, resulting in substantial improvement in the management of public data [24]. The interviewees agree that quality data is a fundamental necessity for implementing AI solutions. Data collection methods vary from one ministry to the other, and also at a departmental level. However, ministries are aiming for paperless processes and digitized systems to increase efficiency.

Interviewee 05

Any process which was paper-based, now it is fully digital.

Yet, there remain some departments whose data collection is done manually via printed forms. As a number of interviewees suggest, the application of AI processes can be severely limited due to the manual handling of data within the particular departments.

The survey respondents were questioned on the format that data was received at their end; including printed hard copies and via digital platforms. Overall, 48% of survey participants receive data online via digital platforms, while 47% receive a mixture of printed hard copies and online forms (Fig. 11.8). Out of 470 respondents who receive data in a digital format, 84% agreed that collecting data this way facilitates a swifter process at work and helps them do their job faster. Nevertheless, unstructured and quality of data remain a major concern for the interviewees, as both are weaving components to assure reliable AI predictions and outcomes.

Interviewee 01

... we do not have the correct data structures, we do not have the correct data quality—data quality is as in availability, as in semantic correctness, as in error free data.

Hence, data needs to be reformatted and sanitized to implement AI solutions. Additionally, the garbage-in garbage-out mantra needs to be reinforced and strictly dealt with once a data sanitization process is applied to the information that will be fed to AI systems. It was discussed that a nationwide data cleanse will allow for a common platform among ministries, however, this process is deemed as a tough challenge, both in terms of preparation and implementation. Moreover, limited use of the available structured data is currently being made, due to the lack of either expertise or human resources that could analytically harness the information and apply it accordingly.

In this regard, the Malta AI Strategy [12–18] establishes that an ecosystem infrastructure will be in place, allowing for investment in data centres, access to supercomputing processes and open data. A framework to explore methods for combining data from sensor-based sources and other relevant data into national open data systems, will also be developed. The effort will involve an investigation of the present open data registry's readiness to foster the growth of AI-enabled apps. However, the interviewees argue that this high-level data vision requires more information on a cross-governmental standardization of data format, structure, collection, sanitization and processing.

Considering that the data (which is already available in different formats) will be fuelling the Government's AI solutions, such internal systems should foster inter-ministerial collaborations to achieve successful results. This is particularly important in view of the adoption of the 'Once-Only Principle', allowing public entities to share citizen data, which is entered once [25]. Raw data necessitates different computing methods and despite AI processes seeming simple, the underlying data structure is frequently the missing element in the development of efficient and reliable AI analytics [25]. Hence, the importance of having standardized data applied across the Maltese government is crucial, as this will allow for reliability, accountability and scalability as and when required.

11.9.4.2 Data Processing

Data processing incorporates the legal framework and ethics related to the processing of data. The interviewees argue that even though it is beneficial to have legal frameworks in place to ensure data sharing, these will also slow processes due to compliance and due diligence procedures. However, this will ensure accountability and guarantee that systems are ethically designed from the beginning. It is also hoped that with such measures in place, organisations as well as individuals will trust these innovations while also attracting foreign investment due to the certification framework in place.

> Interviewee 17
>
> ... it's going to have a very rigid, strict approach to which I think is terrible. I think it's going to stifle innovation, but it's good for Malta because we have a regulatory framework and we're going to attract more business because we're ready.

In this regard, the interviewees were concerned as even though the Ethical AI Framework [12–18] provides direction, the Maltese legislation needs to be updated to be congruent with these innovative undertakings, in order to complement each other rather than operating as two separate legal processes. Furthermore, a total of 55% of the public employees that participated in the survey agree that the public will feel sceptical about using AI systems (Fig. 11.57), while a total of 90% believe that the public needs more information about AI, in order to trust new processes (Fig. 11.58). This shows the need for further information among the general public, which should not only be limited to explain the Government's AI projects, but also describing, in a relative manner, the data processing procedures, thus fostering trust through transparency.

The Government should aim to develop a realistic model to give direction and facilitate AI practices. As outlined in the national strategy, 'Malta Towards Trustworthy AI' policy document [12–18], the Government intends to assist AI practitioners in recognizing and addressing possible risks associated with AI, while also incorporating superior standards into AI solutions. The strategy addresses ethical undertakings in every stage of AI design, development, implementation and provision of results. Additionally, the Malta Digital Innovation Authority (MDIA) will establish a National Technology Ethics Committee to monitor the implementation of the Ethical AI Framework [12–18] and its conjunction with other established protocols. In this regard, it is not defined how the monitoring of ethical process will be implemented by the committee and further information is required. Publicizing the role of the National Technology Ethics Committee, the technology that falls under its remit and the monitoring processes that will be established, will further foster transparency among all the stakeholders.

The IBM Watson team state that for AI to be trustworthy, it has to be human-centered [26]. However, as [27] discuss, changes can cause confusion and uneasiness among the public and insecurity arises from the loss of control associated with autonomous AI. As AI systems process and learn through the data input by humans whereby errors are possible, AI decision making may be biased and present a risk to human beings [28]. Furthermore, [29] argue that despite the efforts to incorporate ethical consideration as part of AI solutions, ethical philosophies as practiced by human law might attest a challenge to define and design in a computable manner. Thus, the importance to fuse ethical processes with AI solutions from the desigining stage is necessary to ensure that a human-centric approach is applied from the early stages.

11.9.5 Preparedness

This section discusses the overarching theme, 'Preparedness', which analyses AI-readiness from the interviewees' point of view, technicalities, availability of funds and competitivity.

11.9.5.1 Ready or Not

The majority of interviewees believe that their ministry is not ready to adopt and implement AI for a number of reasons. Some were absolute in their reply, while others mentioned that their ministry is positively addressing innovation.

> Interviewee 01
>
> The ministry is not ready, it is decades from being ready. There's the wrong mentality, it is the wrong culture.
>
> Interviewee 10
>
> ... ministry is on the right track to be ready.

A number of interviewees confirmed that the extent of readiness varied between departments. They also emphasized how change needed time and that resistance would most likely affect the process. Additionally, they argued that although AI is a central topic for discussion and implementation, the Maltese government's system is not yet ready for AI adoption.

> Interviewee 13
>
> This is not a magic wand situation, as it is very complicated and the Governments' system is not ready for AI.

The Government lacks the necessary structures that help facilitate inter-ministerial AI solutions, resulting from old-fashioned ways of processes, as well as the rigidity of the structure itself. During the last years, the public administration has undergone substantial change in the way service is delivered, especially on the digital side. However, the public service composition has remained unaltered.

The interviewees acknowledge that the situation is complex as change does not happen overnight and AI adoption will affect individuals at all levels. From the interviewees' viewpoint, it seems that resistance is ingrained into the mentality of some public officers, as things have always been done in a certain way, hence change is not necessary. Effectually, the interviewees classified the personnel in two categories; either as early adopters or those who wait, which is humane enough considering the perceptions and misconceptions about AI among the public.

"There are always the early adopters, champions I call them and those who prefer to wait, which is normal. So, you normally have a mixture of different people. Initially we focus on the early adopters, but those are the people who will sell your product. If you enjoy the experience for sure, you're going to talk about it. So other people will start using it. It's a whole process. The challenge with technology is always changing the culture." Interviewee 02.

In addition, the public officers might feel threatened due to the lack of internal strategic action plans, lack of communication and lack of information. Furthermore, as discussed by Dwivedi et al. [30], long-term strategies do not work with quickly evolving technologies, such as AI. In turn, Dwivedi et al. [30] propose building adaptable short-to-medium-term AI strategy plans that can adapt to technological advancements and changes. Putting this notion into perspective, short term ministerial strategies might prove to be more effective as change would be addresses at a proportional rate.

On the other hand, the survey results attest that, as opposed to the interviewees inclination into thinking that the employees are not ready, the survey responses reveal that the public administration is ready to take on the AI challenge. In fact, 91% of the survey respondents are interested to learn how AI can help them perform better at work (Fig. 11.23), while 89% would like to learn about uses of AI (Fig. 11.22). Moreover, 78% think it would be very interesting if AI was introduced at their work (Fig. 11.21), and 58% are confident that learning how to use AI will not be difficult (Fig. 11.19).

11.9.5.2 Technicalities, Funds and Competitivity

Technology-wise, two (2) interviewees agreed that the hardware and software infrastructure to adopt AI is in place, however, opinion varied among the rest.

Interviewee 15

Technically I believe we are. We have the capabilities of pushing the technology as much as the business owner wants from the technology readiness...

Interviewee 14

> I can say no. In fact, to adopt such systems solutions, I had to go for a third-party service. We had to deploy the services. We are still deploying third party services because the technicalities for such solutions cannot be taken onboard by my staff definitely.

Moreover, concern about the agility of the procurement process emerged as such technologies are upgraded or enhanced frequently. It is believed that the current procurement processes hinder the acquisition of the latest systems since by the time of award, newer solutions would have been released. It was discussed that many a time, budgets, estimation of expenses as well as the recurring costs to upkeep the systems are established earlier. However, in terms of competitivity, the Government is restricted, especially in terms of employee engagement protocols and salary packages associated with the respective grades. Hence, the human resources turnover for certain positions is high due to uncompetitive remuneration packages.

> Interviewee 04
>
> In terms of management, in terms of employee satisfaction and in terms of how is the organisation and the employee benefiting realistically, we are losing a number of employees within the public service because of we cannot compete with the private sector or even with the entities as the Government salary scale is lower than the market competition.

The survey did not address funds and competitivity, however, 65% of the survey respondents are undecided or do not know whether the infrastructure to run AI systems is already in place (Fig. 11.31). Nonetheless, in a study published in 2018 by Capgemini Consulting, Malta ranked 19th out of 35 countries, on the AI readiness benchmark, defined on three (3) levels; institutional environment, technological maturity and skills advancement. Moreover, Malta scored highest on 'IT Maturity', which is translated in having in place a digital infrastructure that facilitates the integration of AI solutions. The study further assessed the use of information technology in the country [31]. This aligns with the e-Government Benchmark reports, issued yearly by the European Commission whereby Malta is now the frontrunner, leading with an overall score of 97% [32, 33].

11.9.6 Change What?

This section discusses the overarching theme, 'Change what?', which includes governmental practices and what it takes to drive AI within the Maltese public administration.

11.9.6.1 Current Governmental Practices

The shift in mentality from the public officers has been perceived by the interviewees as a major challenge, as it involves a community of people who have been doing the same job in the same way for a long time.

Interviewee 17

> Hindrance is not from a technological point of view, more from a social-political point of view.

According to the National Statistics Office [8] the public administration totals 50,808 employees of which 30,000 are employed with the public service and 20,000 with the public sector [5–7]. As such, the Government's culture needs to be addressed both horizontally and vertically in order for Malta to successfully achieve AI goals within the public administration. In this regard, it is to be noted that if all public officers are engaged in the Government's AI rollout, so will a segment of the general public. This is because public officers are the same segment of the wider public, that can in turn testify the benefits of AI and evangelize about AI at the workplace, with their immediate family members and friends, thus generating a wave of interest among the wider society. Such a positive ripple effect through word of mouth facilitates acceptance of the change, while instigating an advanced kind of curiosity of which benefits would be unprecedented. Within an institutional context, potential challenges when implementing change include an antagonistic culture, political pressures, weak leadership, ineffective reporting systems and unmanaged consequences [34]. Hence, targeting these factors would facilitate compliance with new systems.

On the other hand, and in contrast to the beliefs of the interviewees, the survey reports that 75% of the respondents are ready to use AI technologies in their day to day operations (Fig. 11.26), while 86% agree to support the Government's strategy towards change in implementing AI (Fig. 11.27), although as aforementioned, 75% of the total respondents are not aware of the Malta AI Strategy [12–18]. Moreover, 78% understand that a change in work processes is required (Fig. 11.35) and 82% are ready for any change that might impact their role (Fig. 11.41). This pro-active approach by the public officers is reflected in the Malta AI Strategy [12–18], in which the necessary cultural shifts and change management processes are not directly addressed. However, this remains a major concern for the interviewees.

Interviewee 16

> AI is very vast and I think it's the culture the problem.

Addressing the change challenge early will foster new moral principles and favour change behaviours. An institution's culture optimally conditions the organisation's purpose, establishes renewed standards and guarantees workers' compliance to innovate systems of work while fostering a system for value transmission [35].

Another important change that must occur is the shift in misconceptions revolving around the notion of AI robots taking over the world, as portrayed in sci-fi movies. This increases anxiety among employees, and thus necessitates an increase in awareness among public officers.

Interviewee 10

> ... delusion that AI and machines will one day take over the human race, when in reality, the sole purpose for AI is to support the human race by giving solutions to complex problems that it would take a number of years for the human mind to accomplish.

Increasing awareness generates trust. Consequently, it is important to educate the workforce to ensure that the new modus operandi is a holistic positive undertaking. Overall, 79% of the survey respondents agree on trusting the change induced by AI, if well-informed about the impact that AI will have on their duties (Fig. 11.36). Furthermore, it transpires that only 7% of the respondents are informed on how the ministry will implement AI (Fig. 11.34), which is in turn consistent with the need to inform the general public workforce about AI adoption.

Interviewee 09

... information is key to AI acceptance among the public workforce and the public. I believe the public workforce and the public need assistance to develop ...

To this effect, 82% of the survey respondents agree that information and knowledge sharing on AI would reduce uncertainty on the uptake of AI as part of their role (Fig. 11.33). Hence, the figure of 94% of survey respondents agreeing that communication is essential in getting the public workforce ready for adoption, is no surprise.

As previously mentioned, the promise of a public awareness campaign [12–18], will tackle uncertainty and misconceptions among the public officers while fostering a positive culture change. Moreover, the Government is committed to establishing a technical committee to examine the architecture of solutions implemented within Public Administration, with a special emphasis on the use of AI. However, it is unsure whether the Government will undertake the task to revise its internal structures related to human capital and human capacity related to the grades and associated roles. In this regard, [36] notes that employing better governance structures would further address the immediate changing needs in becoming an AI-powered organisation.

Dissolving silos is also high on the list of challenges for the interviewees. Silos are a barrier to collaboration, data sharing and knowledge transfer.

Interviewee 11

We shouldn't work in silos anymore, this is one of the most disadvantages that the Government, as yet, we still work in silos; although a lot of improvements were done in the last years ... ... seeing the importance of the have to work together, the one government concept, which I did with the project, whereas entities need to collaborate together to achieve this massive project ...

The Government is also working on dismantling the silos approach with the implementation of the Once-Only Principle, whereas all public entities share citizen data, such that inforamtion is entered only once [25]. To this effect, 69% of the survey respondents are expecting AI to increase collaboration (Fig. 11.44).

11.9.7 *To Drive AI*

From the interviews; positivity, leadership and public inclusion emerged as the three (3) drivers that will power AI adoption. A positive approach to AI adoption will reassure the public servants. This will be further reinforced if leaders are well informed and lead the change in a constructive manner, whereby communication is a frequent process. Furthermore, key players in the adoption of AI technologies within the different ministries or AI champions, will induce curiosity and interest among their colleagues.

Interviewee 04

> The ministry is ready. We are using positive concepts in terms of changing culture. We inform our workforce in terms that going digital is not changing the current processes, but facilitating their workflow.

Public inclusion is also perceived as a significant factor in adopting AI and the interviewees agree that the public should be educated about AI, informed about the projects undertaken by the Government, how their data is being processed and their rights in this regard. In fact, the Malta AI Strategy [12–18] states that the public will be informed about AI and its importance through a nationwide campaign. This will ensure that those most vulnerable understand that AI solutions are also available for their comfort and safety. In this regard, creating a humane share value approach, whereby people can relate to the changes that are happening around them, in a constructive and informative way, will foster a consented acceptance towards the fast approaching AI revolution. The term humane in this context translates to an understanding and empathetic manner; operating from an emotionally intelligent perspective when expecting people to adjust and accept change.

11.9.8 *In Demand*

This section discusses the overarching theme, 'In Demand', which explores knowledge and expertise availability within the Government as well as the education side to AI readiness.

11.9.8.1 Knowledge and Expertise

Interviewees agree that in general, with regard to the implementation of the Malta AI Strategy [12–18], the required technical knowledge and expertise are missing from within the ministries. Additionally, the existing technical knowledge is limited to the few, hence the need to train and upskill is amplified.

Interviewee 09

> As an IMU we do not have the expertise.

Moreover, apart from the workforce, the required training should also be directed towards the business owners. In this regard, the latter are to identify processes that could become AI-enabled, while the interviewees are entrusted with the responsibility to execute such AI endeavours within their ministry [12–18]. However, due to the lack of knowledge and personnel, the offices of the interviewees are burdened with tasks beyond their capacity.

The Information and Communications Technology (ICT) class structure, as defined by the Government, comprises of eight (8) roles with job descriptions, and eligibility rules used to engage ICT professionals within the various ministries' Information Management Units (IMUs). The interviewees agree that this structure is restrictive in terms of engaging the required expertise to meet the emerging needs of IMUs.

Interviewee 14

> The ICT class needs more human resources, with specific expertise to meet the needs of the ever-increasing demand within the ministry.

Furthermore, human resources turnover is high and engaging people is becoming more challenging in this sector. At this stage, the ICT class necessitates technically knowledgeable and specialized skills of which are very limited within the Government. Hence, the need to outsource is customary practice within the public administration.

Interviewee 08

> ... because we don't have resources, we outsource ... engage a third-party assistance in terms of data engineering ...

Nevertheless, restructuring the ICT class by allowing flexibility in their capacity building will foster knowledge sharing while enhancing the skillset within the ministries. A number of interviewees confirm that they are tapping into the resources of the Malta Information Technology Agency (MITA) for direction and assistance on AI-related matter. However, the benefits of having expertise in-house is deemed necessary for long-term AI projects.

When asked whether the ministry had the required technical expertise to implement AI-based systems, 36% of the survey respondents were undecided and 22% did not know (Fig. 11.29). Furthermore, when asked if personnel within the ministry could help implement AI, 34% were undecided and 26% did not know, while 10% disagreed and 29% agreed (Fig. 11.37). On the contrary, 73% consented that third-party experts can help implement AI (Fig. 11.38) and 77% agree that assistance is needed in identifying processes that can be AI-enabled (Fig. 11.28).

11.9.8.2 Educational Movement

Concerns are growing that continued progress and adoption of AI may result in job losses as tasks and maybe entire jobs become increasingly automated [37, 38]. To assure job retention, the Government is embarking on a nationwide reskilling and upskilling campaign encouraging new career paths and especially the creation of new roles. As a result, individuals with a knowledge of AI will advance quicker and as a matter of fact, certain job replacements will occur, especially where repetitive duties are involved. To this effect, interviewees support the specialization and upskilling of the current ICT class, in order to retain that skillset and AI-related knowledge in-house.

Interviewee 11

> Another aspect which needs to improve as a government as a whole is human resources in specialised fields, or training must be provided to identify the individuals who will be implementing AI projects.

The Government agencies are a key source for knowledge and expertise, since they centrally hold AI-related knowledge, while also providing direction in designing and developing proof of concepts. Furthermore, the experts agree that education is a key enabler to the successful adoption of AI solutions in Malta. In order to raise AI awareness and advocates, knowledge of AI should become a common skillset amongst future generations. In line with the Malta AI Strategy [12–18], a number of interviewees advocate the integration of AI concepts in the educational curriculum from an early stage. This will enable the timely identification of those champions that excel in AI, thus channeling them to drive the nation's ambitious targets. In fact, various educational institutions already offer AI-related courses.

Interviewee 13

> ... education system, I would introduce one topic at a time such as process automation, through robotic courses that facilitates for fun learning, such as making a toy move ... potentially encourage those students who are more inclined for these types of jobs and skills, engineers, scientists, programmers.

Education and training also play a critical role in job retention and security. The survey asked a number of questions in this regard. More than half of the respondents (52%) do not feel threatened by AI solutions (Fig. 11.53), as public employment offered security, while 64% state that their job will not be replaced by AI (Fig. 11.54). Moreover, 56% agree that new jobs will be created within the Government (Fig. 11.60), while 85% concur that public officers will need to be retrained (Fig. 11.61). Overall, 58% confirm that some jobs will be replaced (Fig. 11.62). These figures tally with the Chief Information Officers (CIOs) concerns; in fact, 82% agree that AI adoption will be successful if AI topics are introduced early in the learning journey of Maltese students.

In fact, the educational plan is extensive and well laid out in the Malta AI Strategy [12–18]. The Government is committed to carrying out a study to identify those roles and skills that are most susceptible to AI adoption to set in place a sound transition plan and promote training for new skill sets. The public service will also benefit from specialized training sessions to keep abreast with new technologies while ensuring that transferable skills are being acquired. Within the education sector, several initiatives are planned to involve students in AI activities and to enhance the educators' development of AI. Ultimately, AI should become part of the student's compulsory studies, to ensure that basic knowledge is nurtured early thus increasing the uptake of AI-related courses. In this regard, a position paper discussing *The Future of Education and Skills*, published by OECD [39], discussed that for the individual and societal future well-being, broader educational objectives should be established. Unless science and technology progress with a purpose, they may aggravate disparities, exacerbate social instability, and expedite resource depletion. Education is critical in fostering the development of the knowledge, skills, attitudes, and values necessary for individuals to participate in and prosper from an inclusive and sustainable future.

11.9.9 AI as an Asset

This section analyses the overarching theme, 'AI as an Asset', which addresses the benefits of AI across five (5) themes; including enhancing processes, effective use of resources, evidence-based management, investment and challenges.

All interviewees consider AI as an asset and identify numerous benefits of AI solutions (Appendix K). In this regard, all CIOs are eager to have AI systems in place and to reap benefits at an unprecedented rate.

> Interviewee 01
>
> We will automate all that process and the impact will substantially result in efficacy gains of 30% to 70% and reduction of errors by 30% to 70%.

Predictive analysis will assist in evidence-based policy drafting and decision making. Interviewees also agreed that having a robust AI strategy, an ethical framework and the first certification programme attesting that AI solutions are being developed in an ethically aligned, transparent and socially responsible manner, is attractive to foreign investors. However, a concerned interviewee argues that to have such processes in place is time-consuming and as already discussed, the lack of resources, lengthy and inadequate procurement process and the lack of knowledge will hugely impact adoption and progress within the public service.

A total of 79% of survey participants agree that AI provides data for informed decisions (Fig. 11.45), 62% agree that AI is more reliable and consistent (Fig. 11.46) and 81% concur that AI increases speed in working operations (Fig. 11.48). Furthermore, 70% agree that AI will facilitate the roles of the public administrators (Fig. 11.49), while 72% affirm that the needs of the public will be better addressed (Fig. 11.40). This is in line with the Malta AI Strategy [12–18], which seeks to guarantee that the advantages generated by the influx of AI innovation will be accessible to the entire nation and the World Economic Forum (WEF), which invites people to embrace AI as its long-term growth will severely impact society's functions and economy [40]. Thus, interestingly, amid the evident lack of information among the public workforce, the survey respondents could still perceive the benefits AI will generate.

11.9.10 Addressing Risks

Misuse of funds, mal-intent of AI solutions, risk mitigation, identifying accountability, data quality and discontinuation of activities were factors identified during the interviews as foreseen risks related to AI adoption. Proof of concepts are being adopted to mitigate implementation risks, however data quality ranks high on the interviewees concerns. The importance of assuming responsibility was also discussed, and consequently they feel more inclined to leave the decision making to humans, rather than machines. Continuity of AI projects should not be a risk but rather warranted, however, AI technologies come with a hefty price tag and therein lies the possibility of discontinuity.

Interviewee 16

> Another risk is the discontinuity due to lack of funds, disinterest or not enough vision to escalate the idea as individuals are focused on their own work or have different priorities.

Risk is addressed from different perspectives in the Malta AI Strategy [12–18], including cybersecurity, the ethical aspects, as well as the shift in jobs and job losses. Risk management is a prerequisite, and in order to target risk perils, the MDIA, the National Technology Ethics Committee and a further Technical Committee are established to assist in mitigating risks on all levels, such as infrastructural or related to implementation. The education campaigns that are forecasted will also mitigate the risk of lack of awareness and diffused misconceptions on AI-related matters among the public, while empowering the workforce to proactively get interested in AI-enabled work processes.

AI may be used to enhance human's life and lifestyles, therefore generating possibilities; it can also be underutilized, thereby causing economic costs, however, being overutilized and misused fosters threats. Anxiety, lack of information and misplaced fears can all contribute to a society under-utilising AI technology to its full potential [41]. By having an integrated system in place to weave AI solutions within society, the Government will ensure that risks are minimized or efficiently mitigated.

11.9.11 Future Applications of AI

This section analyses the overarching theme, 'Future Applications of AI', which discusses the projects the interviewees are envisaging to implement in the near future.

Citizen-centric AI incentives are at the forefront of technologies to be implemented across government.

Interviewee 18

> driving forward initiatives that citizens can tangibly feel the positive changes AI will facilitate and I do that by working upon activities that are closest to the people.

Another important application of AI is through image processing techniques, which are in turn related to the role of the public officers, to enhance their processes or access reliable information. Predictive analysis will assist in making informed and evidence-based decisions. Chatbots which simulate human-like conversations using natural language processing (NLP) are already being used in a number of ministries and further adoption is envisaged. Hence, the need to have quality data structures in place to start implementing these projects effectively, is imminent. As a matter of fact, the six (6) identified high-profile pilot projects included in the Malta AI Strategy [12–18], have a citizen-centric approach, whereby once implemented, the Maltese citizens shall immediately reap their benefits.

11.9.12 Moving Forward

This section discusses the overarching theme, 'Moving Forward', which analyses the concepts the interviewees deem important for the adoption of AI technologies.

Discussion during the interviews revolved around the necessity for a needs' analysis within the respective ministries, to address both the needs of the officers as well as those of the public.

Interviewee 10

An in-depth analysis is certainly required to address some of the issues ...

This exercise should be carried out to identify the most adequate AI processes and their respective feasibility. By starting small with successful implementations, employee onboarding is more likely to occur. A committee representing the public administration, the private sector and academia could carry out this intra-government need analysis, whose task would be to identify investment and generation of savings through the implementation of AI technologies. One particular interviewee suggests a central entity, which delegates to ministries in this AI endeavour, in order to establish a cross-governmental holistic approach. Furthermore, according to the interviewees, the need for an information campaign is overdue. The list of benefits that such a campaign can ignite is limitless, as it will target the whole of the population, kindling a thirst for knowledge. This must also showcase successful projects, educate adults and have AI knowledgeable individuals reach out into our schools to explain and acquaint students with the upcoming future opportunities and risks. Such an approach will holistically captivate the interest of the public while paving the way for the future of Maltese youth. Interviewees agreed that communication is key to promoting change, hence setting up a clear communication strategy will support a positive change process through a humane approach.

Education was further discussed on various levels within the public administration itself. Concerning the public and youth, information can be delivered through the various educational institutions. As for the public administration, the interviewees are lacking the human capital that is required to deploy AI technologies. The

Organisation for Economic Co-operation and Development (OECD) refers to human capital as those individuals or people whose skills, competencies, knowledge and other characteristics bring economic value to an institution or a country [42]. Hence, the Government should incentivize the ICT class to uptake specialized courses so that IMUs will transform into a body of knowledge that would meet the changing needs of their roles. In addition, a reform of the ICT class will allow the interviewees the required flexibility to build capacity according to the demands of the ministries, hence addressing the human capital issues.

Interviewee 16

> Investing in people who have the right skills and retain resources at the Government level ... Collaboration with the private sector, industrial partners and academic institutions is a key element when learning and implementing.

The aforementioned is also confirmed by the National Employee Skills Survey NESS [43], which identified skills gaps, whilst gaining an understanding of the supply and demand, in various areas of the Maltese labour market. Furthermore, the survey mentions the acquisition of new competencies as a critical component in closing the skills gap and reducing skill mismatches. Concerning digital technologies, the latest ICT Skills Demand and Supply Monitor [44, 45] reveals that the demand for ICT professionals is outpacing the local ICT student population. Representatives of tertiary education providers interviewed voiced worry about dwindling interest in ICT-related subjects. Furthermore, the majority of employers believe that ICT graduates require further training to satisfy the organisation's demands. Due to the scarcity of specialized professionals in certain fields, employers are increasingly assigning such responsibilities to the current employees who are normally assigned to other duties.

Moreover, increased collaboration is needed to have the relevant stakeholders on-board during the early stages. Collaboration via EU projects also serves as a learning experience on best practices and provides information on what other EU partners are doing. Furthermore, it was mentioned that the University of Malta should be more involved in AI undertakings and the Government should engage with local experts for consultancy:

Interviewee 03

> In its thousands of research studies conducted every year, the university should address real life problems when conducting research. It can be a collective effort! The Government should trust the University of Malta with the research needed and not look elsewhere through external consultancy. The university has all the experts needed that can direct and give input while studies should be implemented and not end up on a shelf. The Maltese citizens should have an input on the future of the country and these studies can provide valuable insight

Additionally, other officially appointed bodies, public–private partnerships and social partners are also involved in the run to adopt AI, however, it is noted that except for MITA, the entities are mainly engaged with the private sector. The appointed authorities to handle Malta's innovative endeavours are:

- Malta Information Technology Agency [46], whose role is to administer IT processes government-wide. To date, besides providing the Government with ICT infrastructures, systems and services utilising contemporary technology platforms, MITA also integrates technological advancements in concrete and tangible business solutions assuring the Government's plans and projects are skillfully implemented and the return on investment is maximised;
- Malta Digital Innovation Authority [47, 48], is set up to promote the Government's innovation protocols and impose certification standards to ensure compliance with the national regulatory framework;
- Tech.mt [49], whose aim is to promote innovation, seek talent and assist in market research while creating a forum for local IT businesses to exhibit their work and celebrate their accomplishments;
- Malta [50] is the Economic Development Agency of the Maltese Government that seeks to bring foreign investment opportunities to Malta.

From the Malta AI Strategy [12–18], it stemmed that the Government will also set up a:

- Think-Tank to devise an action plan concerning the implications that autonomous innovations will have on the Maltese employment market;
- Technology Regulation Advisory Committee to collaborate with the MDIA in establishing a legal and operational structure for the Regulatory Sandbox;
- Office of the Information and Data Protection Commissioner (IDPC) will have in place a Data Sandbox to assist organisations that require the usage of personal data in the development or testing of innovative solutions related to AI;
- National Technology Ethics Committee to supervise the Ethical AI Framework [12–18] and its confluence with diverse policy efforts;
- Technical Committee to assess the infrastructure of innovative systems to be deployed within the Public Administration, with special emphasis on the use of AI.

Having such structures in place should ensure that AI technologies are designed, developed and deployed from beginning to end in line with best-practices and using the highest standards. This necessitates of thorough evaluation and monitoring as a minute mishap might carry major consequences for the Maltese government.

Through the survey, the public officers were also asked several questions concerning the envisaged way forward. The majority of the respondents agree that the adoption of AI technologies within the ministries will be successful if:

- There is increased awareness of AI among the public and public workforce (89%)
- The public workforce is informed and trained with the right skills in the use of AI (72%)
- The public is informed about the Government's AI initiatives, and how these will improve the Maltese public service (69%)
- AI is introduced as part of the Maltese education system early to have an AI knowledgeable future workforce (82%)

- Preparation and planning for the introduction of AI is done strategically and for the long term (87%)
- The implementation of AI systems is set against a robust legislative framework to protect data privacy while promoting transparency and accountability (81%)
- The Government starts looking into the possible future economic risks related to AI adoption that might impact the Maltese population (79%).

To this extent, the survey respondents are consistent with the interviewees' views on the best way to implement AI in government. Education is key to cater for the emerging employment needs and trends, since the demand for ICT skills is higher than what the current Maltese education system is supplying. In addition, a strategic and mindful approach to AI implementation is favourable, as it assures the public workforce that their rights and worries are being addressed.

11.9.13 Conclusion

The triangulation analysis presented in this chapter has deepened and widened the understanding of the research context and consequently the questions posed. This revealed a varying degree of both agreement and disagreement on the different themes, between the data resulting from the interviews and the survey. The next chapter details the conclusions inferred from this research, a set of recommendations, and suggestions for further research and practice.

11.10 Conclusions and Recommendations

This research investigated the extent of Artificial Intelligence (AI) readiness in the Maltese public administration (main research question) and consequently the necessary change required for the successful implementation, in view of the introduction of AI-related technologies (sub-question). The study applied a mixed-methods approach, whereby semi-structured interviews were conducted with fourteen (14) Chief Information Officers (CIOs) within the public service and four (4) Chief Information Technologists (ICTs) engaged within the public sector. Following the analysis of the qualitative research, survey questions were drawn up and disseminated amongst the entire public administration (population of 50,808). The outcomes from the survey lead to a triangulation analysis, whereby the interview results were validated vis-à-vis the 494 survey responses, the Malta AI Strategy [12–18] and related literature.

This chapter summarizes the major research findings in relation to the study's objectives and research questions, as well as suggesting areas for future investigation. Based on these findings, a set of recommendations addressing the necessary change required for the introduction of AI-related technologies in the public administration, will be drawn for consideration.

11.10.1 Conclusions

The findings from this research are the first kind of detailed contribution towards the introduction of AI in the Maltese public administration. The triangulation analysis presented previously, gives rise to a number of important conclusions with regard to the extent of AI readiness in the Maltese public administration. These have been categorised thematically and underpin a set of recommendations in view of the successful introduction of AI-related technologies in the public administration.

(i) Awareness among the public administration and general public

Both the interviewees and the survey respondents deem awareness of AI and AI-enabled systems and processes to be key in diffusing misconceptions, increasing recognition of AI solutions and minimising the skills gap disparities. This will enable the adoption of AI technology to its full potential [41]. Throughout the evaluation of the survey responses, there is a comprehensive understanding that the majority of the public administration workforce can relate to some form of AI use, are forward-looking at adopting such technologies and believe that AI can simplify work-related processes. What is striking though is that although 86% show a positive attitude towards acceptance of AI-related procedures, as well as the wilfulness to be involved and informed about AI during this transitioning process, 75% of the overall respondents were not knowledgeable of the Malta AI Strategy [12–18]. This is further corroborated by the lack of information about AI and its adoption, reported by 68% of the respondents. Moreover, 90% agree that the general public needs more information about these technologies in order to trust the new processes. These findings indicate that whilst the public administration is very much open to the use of AI, further information and awareness are required, as these are two important precursors for technological adoption to occur [51], both within the public administration and the among the general public. This is in line with the vision of the Government, which aims to launch awareness campaigns targeted at public officers and the society in general. These will provide the necessary information, build knowledge and showcase the benefits of AI, as per the Malta AI Strategy [12–18]. Thus, the research findings, further corroborate the need for such awareness campaigns, which will be instrumental in the successful adoption and implementation of AI in the public administration.

(ii) Education, upskilling and reskilling

The research findings highlight the key role that education plays in view of the introduction of AI-related technologies in the public administration. This will enable the foundations for the required human capital to address the challenges and capitalise on the opportunities [52]. Furthermore, the study reveals that education is critical not just in terms of students in compulsory, further and higher learning institutions, but also within the public administration (in terms of the required upskilling and reskilling). In fact, both interviewees and survey respondents agree to need of the development of different career paths, combined with training, reskilling and upskilling of the public workforce. Moreover, 82% of the survey respondents believe that early introduction of AI-related subjects as part of the Maltese education system, will facilitate the creation of an AI-knowledgeable future workforce. This is strongly reiterated throughout the Malta AI Strategy [12–18], which recognises the central function of education in strengthening the country's knowledge both within the public workforce and across all educational institutions.

(iii) Organisational Reforms

The findings of this research, resulting from the survey data and the interviews with the CIOs and ICTs, provide insights into a number of important organisational reforms, namely related to the:

(a) Data management

The adoption of AI significantly exceeds the boundaries of existing data processing and analytical capability, resulting in substantial improvement in the management of public data [24]. The interviewees agree that quality data is a fundamental necessity for implementing AI solutions. However, they argue that a cross-governmental standardization of data format, structure, collection, sanitization and processing is required, as also attested by the survey respondents. In fact, 30% of the respondents agree that the current data collection processes could hinder AI adoption, while a further 37% are undecided. This could be due to the legacy systems which are not adequate for the changing needs and future requirements of the different departments. This is even more important in view of the upcoming adoption of the 'Once-Only Principle', allowing public entities to share citizen data, which is entered in the system once [25]. The AI-readiness of the data has also raised important concerns in terms of AI-powered data processing and the resulting ethical considerations. As [27] suggest, the loss of control associated with autonomous AI data processing, can potentially cause confusion and uneasiness. Furthermore, as AI systems process and learn through the data input by humans whereby errors are possible, AI decision making may be biased and present a risk to human beings [28]. To this effect, Mitterlstadt [29] argue that despite the efforts to incorporate ethical consideration as part of AI solutions, ethical philosophies as practiced by human law might attest a challenge to define and design in a computable manner. Thus, the importance to fuse ethical processes with AI solutions from the designing stage is necessary to ensure that a human-centric approach is applied from the early stages, with regard to the format, structure, collection, sanitization and processing of data.

(b) ICT (Information and Communication Technology) class

Through the interviews, it also transpired that the lack of human capital within the ministries is persistent, both horizontally and vertically. This is in line with the latest ICT Skills Demand and Supply Monitor Report for Malta [44, 45], which confirms that notwithstanding the interest in AI and other emerging technologies, the uptake and adoption remain low due to lack of specialised resources. Whilst this could be addressed through the provision of expert training and targeted re-skilling, it is also noted that the grading framework implemented by the Government is rigid by way of remuneration packages offered, which limits the engagement of knowledgeable personnel due to lack of financial competitivity. Furthermore, the ICT class is disrupted by the inflexible system on which the department operates, limiting the possibility for long-term advances to meet modern-day emerging needs. As such, reliance on expertise from the private sector is undisputed. While in itself, this is a collaborative undertaking, in the long term it dissipates the notion of knowledge retention which the ministries direly need. Furthermore, as Cath et al. [53] argue, the Government's central role in social and political accountability and long-term nationwide planning is crucial for the fair sharing of benefits and prospects of AI adoption for all citizens.

(c) Centralisation of AI-related expertise

The study has also raised important considerations with regard to the centralization of knowledge. The interviewees suggest the setting up of a central public service department which acts as the AI governmental body to assist and guide the different departments in their endeavour to transform their respective ministries into AI-organisations. To this extent, the CIOs propose the establishment of a common intra-government structure, which is capable of integrating common procedures while also fostering synergies. This should comprise of in-house experts in various AI areas; ranging from data scientists and engineers to software developers. It should serve as an AI knowledge repository, whereby technical knowledge sharing and transfer are the predominant organisational activities. Such initiative would abate the silos concept, which was highlighted by both the interviewees and survey respondents, while advocating collaboration and transfer of knowledge among ministries. To this extent, a collaborative approach to governance will build the government's resilience, while fostering unity in resolving risks and addressing concerning matters [54]. However, as literature suggests, such an endeavour should be well-designed in terms of both its scope and mandate, while ensuring that a politically-accepted model of monitoring, coordination and evaluation is in place [55].

(iv) Collaboration with the University of Malta (UM)

The UM's expertise is derived from the immediate availability of competencies in multiple academic fields, which are valuable to the Government's endeavour in implementing the Malta AI Strategy [12–18]. Additionally, the UM will never grow short of cross-departmental expertise, hence the knowledge that is currently missing within the Government can be supplied through the establishment of a Centre for Applied AI (CAAI) within the UM. As suggested by the interviewees, apart from consulting and assisting the Government, the centre should also take on the responsibility to promote a cross-departmental academic and applied research approach to the implementation of AI technologies in local and international contexts. This is in line with the efforts posed by the Government in the Malta AI Strategy [12–18] aimed at accelerating

AI awareness programmes, whilst fostering collaboration with and within academic institutions.

11.10.2 Recommendations

Underpinned by the aforementioned conclusions and in addressing the necessary change required for the introduction of AI-related technologies in the public administration, the following recommendations have been drawn.

(i) Embark upon awareness campaigns among the public administration and general public

The need for information among the public servants and the general population resounds clearly throughout the research. The public information campaign should target the following:

- What is AI.
- How does AI relate to everyday lifestyle.
- Inform how AI is shaping the public service and its benefits, especially in the citizen-centric activities.
- Inform about the ethical safeguards in place and their purpose.
- Explain the vision of Government vis-à-vis the Malta AI National Strategy and how it intends to achieve it.
- Offer upskilling, reskilling and related training courses.

(ii) Focus on education, upskilling and reskilling

Education is considered as the key factor to success, and this is more accentuated when it comes to AI adoption. The research has raised important considerations for education, upskilling and reskilling within the public administration and across all learning institutions. To this end:

- Career paths should be designed and incentivised to meet the emerging needs of the ministries.
- A cross-ministerial needs analysis is required in order to identify the existing skills gaps and assure job retention.
- AI-related discussions and activities should be introduced in educational institutions from a young age, for AI terminologies to become mainstream. Apart from instigating curiosity among the youngsters, this will further popularise digital career paths.
- AI as an inter-disciplinary educational methodology should be applied in order to ensure that knowledge about AI is transferred across multiple academic disciplines and students can relate to AI in different contexts.

Due to the evolving nature of AI, the area should be considered as one of the main life-long learning themes.

(iii) Implement organisational reforms

A number of important organisational reforms aimed at enabling the introduction and facilitating the implementation of AI in the public administration, should be considered. These include:

- A government-wide strategy for the standardisation of the management of data, including format, structure, collection, sanitization, processing and ethical considerations.
- Rethinking of the current ICT class to gauge interest and allow for flexibility in employing the necessary human capital and required expertise, at par with what private organisations offer in terms of remuneration competitiveness and work-related benefits.
- The setting up of a central public service department which provides the ministries with in-house expertise, technical knowledge and direction for the design, development and application of AI technologies.

(iv) Increase collaboration with the University of Malta

The setting up of a Centre for Applied AI (CAAI) within the UM, whose role would be to:

- Provide the Government with the necessary cross-departmental consultancy related to AI expertise and knowledge, and assist the Government in its AI research efforts.
- Create a national dialogue about the many aspects of AI, therefore advancing a comprehensive discussion about the implications of automation on our labour market and how we might prepare for impending problems.
- Include outreach projects with educational institutions and organise dissemination events for young adults.
- Engage students reading AI-related subjects to target real-life problems when conducting research, while promoting selective AI-research proposals for prototyping in the Maltese society.

11.10.3 Future Research

As a result of the findings from this study, a number of significant implications for all the stakeholders involved in the adoption of AI technologies in the public administration arise. First and foremost, the awareness of AI for both the Government workforce and the general public is paramount. Such recognition and appreciation of the benefits is to be coupled with a solid educational undertaking, across compulsory, further and higher education. Complemented by the upskilling and reskilling of the current labour workforce, such endeavours will pave the way for the human capital, knowledge and expertise, required to address the challenges and exploit the opportunities brought about by this emerging technology. The research findings also suggest that a number of key organisational reforms within the public administration have the potential to enable and facilitate the extent of AI readiness. These include the standardisation of the management of data across ministries, the rethinking of the current ICT class and the establishment of a central public service department which provides the required AI-related technical expertise. Such efforts can be augmented through further collaboration with the UM in adopting a cross-disciplinary approach to applied AI research, whilst assisting the Government in its nationwide endeavours.

Although identified threats to validity and reliability [9] have been addressed by a series of mitigation measures prior to the commencement of the research, the study is still restricted by a number of limitations; including the lack of previous literature on the subject in the local context and issues with the selected populations (as discussed in the Research Methodology chapter). As such, based on the findings, possible implications and recommendations and the identified limitations, further research is recommended. Interviews among the public workforce would complement the current survey responses, whilst giving a more in-depth and contextual understanding of the challenges and benefits of the adoption of AI [56]. These could be complemented by interviews with the private sector and other important stakeholders, such as academia, which as the findings suggest, play an important role in the research context. Finally, a more longitudinal approach to researching the overall introduction of AI in the public administration, its challenges and benefits, will provide an evidence-based underpinning for future policy and practice.

References

1. Watson R (2015) Quantitative research. Nurs Stand 29(31):1–7
2. Shafritz J, Russell EW, Borick C, Hyde A (2016) Introducing public administration, 9th edn. Routledge, NY
3. Government of Malta (2018) The public service and the public sector. https://publicservice.gov.mt/en/Pages/The%20Public%20Service/PublicServicePublicSector.aspx. Accessed 16 June 2021
4. ANSEK (2014) Schedule of grades. http://www.ansek.org.mt/resources/files/Estimates_2014_1403112621.pdf. Accessed 04 Aug 2021
5. Government of Malta (2021a) New public service strategy. https://publicservice.gov.mt/en/pages/New_Public_Service_Strategy.aspx. Accessed 03 Oct 2021
6. Government of Malta (2021b) Press release by the offcie of the principal permanent secretary: Public service week 2021 kickstratrs today. https://www.gov.mt/en/Government/DOI/Press%20Releases/Pages/2021/May/28/pr211019en.aspx. Accessed 12 Aug 2021
7. Government of Malta (2021c) The public service and the public sector. https://publicservice.gov.mt/en/Pages/The%20Public%20Service/PublicServicePublicSector.aspx. Accessed 12 Sept 2021
8. NSO (2021) Registered employment: January–March 2021. National Statistics Office, Malta
9. Surucu L, Maslakci A (2020) Validity and reliability in quantitative research. Bus Manage Stud Int J 8(3):2694–2726
10. Whiston SC (2016) Principles and applications of assessment in counseling, 5th edn. Cenage Learning, CA

11. Jordan C, Hoefer RA (2001) Reliability and validity in quantitative measurement. In: Thyer BA (ed) The handbook of social work research methods. SAGE Publications Inc., Thousand Oakes, pp 53–68
12. Government of Malta (2019a) MALTA the utimate AI launchpad; A strategy and vision for artifical intelligence in Malta 2030. Government of Malta, Valletta
13. Government of Malta (2019b) Malta towards an AI strategy: high level policy document for public consultation. https://malta.ai/wp-content/uploads/2019/04/Draft_Policy_document_-_online_version.pdf. Accessed 12 June 2021
14. Government of Malta (2019c) Malta AI. https://malta.ai/. Accessed 10 Aug 2021
15. Government of Malta (2019d) Malta: the ultimate AI launchpad. https://malta.ai/wp-content/uploads/2019/11/Malta_The_Ultimate_AI_Launchpad_vFinal.pdf. Accessed 21 Aug 2021
16. Government of Malta (2019e) Malta:towards trustworthy AI: Malta's ethical AI framework. MDIA, s.l.
17. Government of Malta (2019f) Mapping tomorrow: a strategic plan for the digital transformation of the public administration 2019–2021. Office of the Principle Permanent Secretary (Office of the Prim Minister) and MITA, Valletta
18. Government of Malta (2019g) Mapping tomorrow; A Strategic plan for the digital transformation fo the public admnistration 2019–2021. Government of Malta, Valletta
19. Nightingale AJ (2020) Triangulation. In: Kobayashi A (ed) International encyclopedia of human geography (2nd edn). Elsevier, s.l., pp 477–480
20. European Commission (2018) A definition of AI. https://ec.europa.eu/futurium/en/system/files/ged/ai_hleg_definition_of_ai_18_december_1.pdf. Accessed 18 Sept 2021
21. Muthukrishnan N et al (2020) Brief history of artificial intelligence. Neuroimaging Clincs North Am J 30(2020):393–399
22. Office for Artificial Intelligence (2020) A guide to using artificial intelligence in the public service. Office for Artificial Intelligence, British Government, London
23. Bipartisan Policy (2020) AI and the workforce. https://bipartisanpolicy.org/download/?file=/wp-content/uploads/2020/07/BPC_AI_Whitepaper_RV3.pdf. Accessed 15 Sept 2021
24. Wirtz BW, Muller WM (2019) An integrated artificial intelligence framework for public management. Public Manag Rev 21(7):1076–1100
25. Krimmer R (2021) The once only principle. https://toop.eu/. Accessed 12 June 2021
26. IBM Watson (2021) Trustworthy AI is human-centered. https://www.ibm.com/watson/trustworthy-ai. Accessed 21 Sept 2021
27. Wirts BW, Weyerer J, Geyer C (2018) Artificial intelligence ad the public sector—addpications and challenges. Int J Public Adm 42(7):596–615
28. Krausova A (2017) Intersections between law and artificial intelligence. Int J Comput 27(1):55–68
29. Mitterlstadt BD et al (2016) The ethics of algorithms: mapping rhe debate. Bug Data Soc 3(2):1–21
30. Dwivedi YK et al (2021) Artificial intelligence (AI): multidisciplinary perspective on emerging challenges, opportunities, and agenda for research, practice and policy. Int J Inf Manage 57(101994):1–47
31. Tinholt D (2018) Artificial intelligence readiness and performance benchmark. https://www.capgemini.com/2018/07/artificial-intelligence-readiness-and-performance-benchmark/. Accessed 08 June 2021
32. European Commission (2020a) eGovernment benchmark 2020: eGovernment that works for the people. https://digital-strategy.ec.europa.eu/en/library/egovernment-benchmark-2020-egovernment-works-people#:~:text=eGovernment%20Benchmark%202020%3A%20eGovernment%20that%20works%20for%20the%20people,-eGovernment%20in%20Europe&text=The%20European%20frontrunners%. Accessed 24 Sept 2021
33. European Commission (2020b) State of the Union addresses by President von der Leyen at the European Parliment. https://ec.europa.eu/commission/presscorner/detail/en/SPEECH_20_1655. Accessed 16 June 2021

34. Mogogole KE, Jokonya O (2018) A conceotual framework for implemeting IT change management in public sectors. Proc Comput Sci 138:835–842
35. Kumar A (2016) Redefined and importance of organisational culture. Glob J Manage Bus Res; Adm Manage 16(4):15–18
36. Zuiderwijk A, Chen YC, Salem F (2021) Implications of the use of artificial intelligence in public governance: a systematic literature review and research agenda. Gov Inf Q 38(101577):1–19
37. Mehr H, Ash F, Fellow D (2017a) Artificial intelligence for citizen services and government. Ash Centre for Democratic Governance & Innovation, Harvard Kennedy School, Volume August, pp 1–12
38. Mehr H, Ash F, Fellow D (2017b) Artificial intelligence for citizen serices and government. Harvard Ash Center Technology & Democracy Fellow, August(1), pp 1–12
39. OECD (2018) The future of education and skills: education 2030. OECD, s.l.
40. Kande M, Sonmez M (2020) Don't fear AI. It will lead to long-term economic growth. https://www.weforum.org/agenda/2020/10/dont-fear-ai-it-will-lead-to-long-term-job-growth/. Accessed 25 Sept 2021
41. Floridi L et al (2018) AI4People—An ethical framework for a good AI society: opportunites, risks, principles and reccommendations. Mind Mach 28:689–707
42. Insights OECD (1964) The vaue of people. OECD, s.l.
43. Jobsplus, NCFHE & Malta Enterprise (2021) National employee skills survey. https://ncfhe.gov.mt/en/research/Documents/Employee%20Skills%20Survey/Employee%20Skills%20Survey%20report.pdf. Accessed 25 Sept 2021
44. eSkills Malta Foundation (2021) ICT skills demand and supplu monitor. https://eskills.org.mt/en/demandsupplymonitor/Documents/. Accessed 04 Sept 2021
45. eSkills Malta Foundation (2021) ICT Skills demand and supply monitor 2021. https://eskills.org.mt/en/demandsupplymonitor/Documents/ICT_Demand_Supply_Monitor_2021.pdf. Accessed 25 Sept 2021
46. MITA (2021) MITA. https://mita.gov.mt/about-us/. Accessed 09 June 2021
47. MDIA (2021a) Malta digital innovation authority. https://mdia.gov.mt/. Accessed 14 July 2021
48. MDIA (2021b) Malta's national AI strategy ranked 10th best amongst 54 countries. https://mdia.gov.mt/author/mdiaadmin/. Accessed 03 July 2021
49. Tech.mt (2021) We promote the Malta digital innovation technology strategy. https://tech.mt/technology-industry-malta/. Accessed 25 Sept 2021
50. Malta Enterprise (2021) Malta enterprise. https://www.maltaenterprise.com/why-malta. Accessed 25 Sept 2021
51. Simtowe F, Muange E, Munyua B, Diagne A (2012) Technology awareness and adoption: the case of imporved pigeonpea varieties in Kenya. https://ageconsearch.umn.edu/record/126760/. Accessed 03 Aug 2021
52. Atiku SO, Boateng F (2019) Rethinking education system for the fourth industrial revolution. In: Human capital formation for the fourth industrial revolution. IGI Global, Hershey, pp 1–17
53. Cath C et al (2018) Artificial intelligence and the 'good society': the US, EU, and UK approach. Sci Eng Ethics 24(2):505–528
54. Bouwer R, Pasquini L, Baudoin MA (2021) Breaking down the silos: Building resiliance through cohesive and collaborative soical networks. Environ Dev 39(100646):1–12
55. Cihon P, Mass MM, Kemp L (2020) Should artificial intelligence governance be centralised? Design lessons from history. In Proceedings of the AAAI/ACM conference on AI, ethics, and society, vol February, pp 228–234
56. Adams WC (2015) Conducting semi-structured interviews. In: Wholey JS, Hatry HP, Newcomer KE (eds) Handbook of practical program evaluation. Wiley & Sons Inc, CA, pp 492–505

Appendix A
Consent form for Interviews

One-to-One Interview Consent Form

- ___________ confirm that I am participating in this research study out of my own free will.
- I am aware that I have been invited to participate in this interview in which the researcher will ask me questions to investigate the AI readiness in the ministry under my care.
- I am aware that my participation in the research study will last approximately an hour.
- I am aware that I can withdraw from the research study at any point with no consequences. In the event that I choose to withdraw from the study, any data collected from me will be erased.
- I am aware that I can refuse to answer any question asked by the researcher with no consequence.
- I am aware that I can choose to have my data withdrawn from the research study at any point with no consequence and that this data will be erased from storage.
- The nature and purpose of the study have been explained to me in writing and I have been given adequate time to read it and ask any questions regarding the research study.
- I know that my participation in this research study involves being asked question relation to the preparation of the ministry vis-à-vis the introduction of AI technologies.
- I am aware that I have the right to access the information I have provided at any time while it is in storage.
- I am aware that there will be no direct benefit to me by participating in this research study.
- I am aware that this research study carries no physical risk as the interviews will be carried out online as well as none of the psychological and socio-economical risks as the interview questions will be solely related to the investigation of AI

M. Sciberras and A. Dingli, *Investigating AI Readiness in the Maltese Public Administration*, Lecture Notes in Networks and Systems 568,
https://doi.org/10.1007/978-3-031-19900-4

readiness in the relevant ministry. Regarding confidentiality risks, the interviews will be numbered, hence no reference to either the ministry or the interviewee will be made available at any time.

- I give permission for my interview to be audio—recorded.
- I am aware that any audio recording will be destroyed once the research study has been completed.
- I am aware that all safeguards will be taken to protect my privacy according to the General Data Protection Regulations (GDPR) and the Malta Data Protection Act 2018. I have the right to access, rectify, and where applicable, ask for the data concerning me to be erased.
- I am aware that all information provided by me will be kept confidential on the researcher's private laptop that is password protected.
- I am aware that my identity will remain anonymous, and personal information will not be revealed in any publications, reports or presentations arising from this research. All the necessary steps to protect my identity will be taken by the researcher.
- I am aware that extracts from my interview may be quoted in the research study, as long as my identity remains protected (my identity will not be noted on transcripts or notes from my interview).
- I am aware that my signed consent form and audio recordings will be kept on the researcher' private laptop that is password protected. Only the researcher and the assigned tutor will have access to the data for a period of two years after the completion of the study.
- I have been provided with information regarding IDEA Leadership and Management Institute (ILMI).
- I have been provided with the contact details of the researcher and I am aware that I can contact the researcher to seek further information.
- I have been provided with a copy of the information letter and understand that I will also be given a copy of this consent form.

Information about Research Participant

____

Signature

____

Full Name (in caps)

Date:

Information about Researcher

____

Ms. MARVIC SCIBERRAS

Date:

Appendix B
Interview Questions

Q1. Define AI in your own words.

Q2. What are some technologies already available in the ministry?

Q3. What are the current methods for data collection, analysis and storage? [1]

- How long have these processes been in place? Are these methods effective? Are they digital? How do you plan to transition from paper to digital?

Q4. Has the ministry a strategy in place to facilitate the introduction of AI? [2]

- Yes: what is the strategy about? What does it target? Does is involve regulating factors to protect information provided and privacy?
- No: how does the ministry envisages the introduction of AI to take place?

Q5. What kind of AI solutions will the ministry be looking at adopting? [3–5]

- Has the ministry made a study to identify the AI technologies needed for its workforce and the public? (machine learning, deep learning for data processing and analysis, conversational AI—chatbots to facilitate communication with public, software components to eliminate repetitive administrative work, robots as ushers or receptionists, etc.)

Q6. Do you think the ministry is ready and in what ways?

- in terms of: transformational readiness—to improve operations, reduce costs and increase efficiency in expenditure [6]
- technical readiness—how similar or different will AI need to be? Is data available for processing/is it already in digital form? From a legal perspective to abide by gdpr? [7]
- organisational readiness—is the introduction supported by management and

M. Sciberras and A. Dingli, *Investigating AI Readiness in the Maltese Public Administration*, Lecture Notes in Networks and Systems 568,
https://doi.org/10.1007/978-3-031-19900-4

public administrators alike? How about the human aspect? Is the workforce prepared/ready/skilled enough? Is a common vision in place? [8]
- environmental readiness—are you aware of other gov entities (local and foreign) already using AI? Is the ministry environment AI friendly? What's the envisaged impact on the work place? Is/how will the strategy be communicated to the employees? [9]
- Has a study been done to calculate the financial needs for the introduction of AI in the ministry?

Q7. In what ways do you think the introduction of ai would benefit your ministry? [10, 11]

- Explain why and how it will strengthen the current operations and processes?

Q8. In your opinion, are there any predicted risks with the introduction of AI? [12]

- Yes: why are these considered as risks? What actions will be taken to minimise or mitigate these risks?
- No: in what terms there is no risk? Does it mean that all the employees will smoothly accept the introduction of AI? What is the ministry doing to avoid risks?

Q9. What changes is the ministry foreseeing regarding the introduction of AI? [13, 14]

- Organisational changes (restructuring)
- Upskilling/redefinition of jobs/designing of new job opportunities (programmers of robots/chatbots, machine learning analysts?

Q10. In your opinion, what would be the best way forwards towards implementing an AI enhanced public service? [15, 16]

- Explain why

Q11. Is there anything else you would like to add?

References

1. Eljasik-Swoboda T, Rathgeber C, Hasenauer R (2019) Assessing technology readiness for artificial intelligence and machine learning based innovations. Data, pp 281–288
2. Andrews W (2017) Applying artificial intelligence to drice business transformation: a Gartner trend insight report, 29: Gartner Inc.
3. Arrietaa AB et al (2020) Explainable aritficial intelligence (XAI): concepts, taxonomies, opportunites and challenges towards responsible AI. Inf Fus 58:82–115

4. Castelo-Branco I, Cruz-Jesus F, Oliviera T (2019) Assessing industry 4.0 readiness in manufacturing: evidence for the European Union. Comput Ind 107:22–32
5. Cihon P, Mass MM, Kemp L (2020) Should artificial intelligence governance be centralised? Design lessons from history. In: Proceedings of the AAAI/ACM conference on AI, ethics, and society, vol February, pp 228–234
6. Yang Z, Ng B-Y, Kankanhalli A, Luen-Yip JW (2012) Workarounds in the use of IS in healthcare: a case study of an electronic medication administration system. Int J Hum Comput Stud 70:43–65
7. Alsheibani S, Cheung Y, Messom C (2018) Artificial intelligence adoption: AI-readiness at firm-level. PACIS, Yokohama
8. Zuiderwijk A, Chen YC, Salem F (2021) Implications of the use of artificial intelligence in public governance: a systematic literature review and research agenda. Gov Inf Q 38(101577):1–19
9. Cath C et al (2018) Artificial intelligence and the 'good society': the US, EU, and UK approach. Sci Eng Ethics 24(2):505–528
10. Bughin J (2018) Marrying artificial intelligence and the sustainable development goals: the global economic impact of AI. https://www.mckinsey.com/mgi/overview/in-the-news/marrying-artificial-intelligence-and-the-sustainable. Accessed 19 July 2021
11. Dignum V (2018) Responsible artificial intelligence. What is artificial intelligence. Springer, NY, pp 9–34
12. Barrett AM, Baum SD (2016) A model of pathways to artificial superintelligence catastrophe for risk and decision analysis. J Exp Theor Artif Intell 29(2):397–414
13. Donald M (2019) Organisational implications. Leading and managing change in the age of disruption and artificial intelligence. Bingley, Emerald Publishing Limited, pp 121–141
14. Yeun SC, Yaoyuneyong G, Johnson E (2011) Augemented reality: an overview and five directions for AR in education. J Educ Technol Dev Exch 4(1):119–140
15. Bouwer R, Pasquini L, Baudoin MA (2021) Breaking down the silos: building resiliance through cohesive and collaborative soical networks. Environ Dev 39(100646):1–12
16. Bughin J, Manyika J (2019) Your AI efforts won't succeed unless they benefit employees. https://www.mckinsey.com/mgi/overview/in-the-news/your-ai-efforts-wont-succeed-unless-they-benefit-employees. Accessed 19 July 2021

Appendix C
Consent Form and Survey Questions

Investigating AI readiness in the Maltese Public Administration

Introduction to the questionnaire

Dear participant,

My name is Marvic Sciberras, and I am currently reading a Master's Degree in Management specialising in Project Management with the IDEA Academy. Pertaining as part of my studies is this research and you are welcome to contact me on sciberras.marvic@gmail.com for any queries you might have related to this questionnaire.

Thank you for your interest in participating. This survey is being compiled with the aim to investigate how ready are the public servants in up-taking and using Artificial Intelligence (A.I.) as part of their job. The survey will help give insightful feedback on whether the public workforce is willing to work with A.I. and identify, what in your opinion, lacks in order to a have an A.I. ready Maltese Public Administration. Upon your feedback, this research will present a set of recommendations to the Maltese Authorities on how best to adopt A.I. in the public administration to meet the workforce's needs.

Please, do note that participation to this questionnaire voluntary and responses are thoroughly confidential and anonymous. All data will be used for academic and research purposes only, and data deletion will take place as per GDPR. Spontaneous responses are encouraged to gather valid data for a beneficial study to A.I. readiness in the public administration.

This questionnaire will only take you 6–8 min to complete and you can withdraw from the questionnaire at any point by clicking the exit link located on the top right-hand side of your monitor.

M. Sciberras and A. Dingli, *Investigating AI Readiness in the Maltese Public Administration*, Lecture Notes in Networks and Systems 568,
https://doi.org/10.1007/978-3-031-19900-4

Your colleagues across the public administration will be responding to this survey too, so thank you for joining; it is sincerely appreciated.

Many thanks for your time and participation

Marvic Sciberras

sciberras.marvic@gmail.com

Demographics

Age 18–25, 26–35, 36–45, 46–55, 56+

Male/Female/Other

Select Ministry Responsible (list all ministries).

Select Present Scale (list all scales 1–19)

Years Employed in Public Administration 0–5, 6–10, 11–15, 16–20, 21–25, 26–30, 31–35, 36–40, 41–45, 45+

Survey Questions

1. **What is Artificial Intelligence?** [17]
   - Blank—write your own answer
   - I am not sure
   - I do not know
2. **Is AI technology already in use in your ministry?** [18, 19]
   - Yes, it is.
     - Blank—give example—mention
   - No, it is not.
   - I am not sure
3. **Are you aware that in 2019, the Maltese Government launched a strategy and vision for AI adoption?** [20–26]
   - Yes, I am aware.
     - Yes—Will this affect your role?

       Yes—How?
No
   - No, I am not aware
4. **Data Collection, Analysis and Processing**

   I. **As part of your role, how do you receive information?** [18, 19]

   (chose as appropriate)

- Printed forms
- Online via digital platforms
- Printed and Online
- I do not receive information
- I do not know

II. **By collecting information this way, I:** [27]

(choose as appropriate—strongly disagree/disagree/undecided/agree/strongly agree)

- Do my job faster
- Slow down as it is time consuming
- Find it difficult to do my job

III. **How do you prefer to receive information?**

(rate your preference—1 being least preferred, 5 being most preferred)

- Prefer information received on paper
- Prefer information received digitally

5. **AI at Work**

(choose as appropriate—strongly disagree/disagree/undecided/agree/strongly agree/I do not know)

- AI can help me simplify and process information quicker [28]
- AI can help me solve problems [29]
- AI can help me take informed decisions [30, 31]
- Through automation, AI can help me enhance my work performance [32]
- AI is better than humans in analysing information [33]
- I will find it difficult to learn how to use AI systems [34]
- I do not need AI to do my job
- It would be very interesting if AI solutions were introduced at work [35]
- I would like to learn more about AI uses [36]
- I would like to learn how AI can help me perform better at work [37]

6. **AI Readiness among the Public Administration Workforce**

(choose as appropriate—strongly disagree/disagree/undecided/agree/strongly agree/I do not know)

I .**Transformational/Strategic Readiness**

- When compared to other technologies, AI is the one that will help me improve my work
- AI can facilitate collaboration with other departments and government agencies [38]
- I am ready to use AI technologies in my day-to-day operations [39]

- I will support the government's strategy towards change in implementing AI [40]
- I need assistance in identifying those processes that can be enabled by AI [41]

II. **Technical Readiness** [42]

- The ministry has the required human capital to implement AI based systems
- The ministry has the required technical knowledge to operate AI based systems
- The infrastructure to run AI systems is already in place

III. **Organisational Readiness** [43]

- Communication is essential in getting the public workforce ready for AI adoption [44]
- Information and knowledge sharing on AI reduces uncertainty regarding the adoption of AI as part of my role
- I am informed on how the ministry will implement AI
- I understand that a change in work processes is required
- I will trust the change AI will bring along if I am informed on the impact AI will have on my job [7]

IV. **Environmental Readiness** [45, 46]

- There are personnel within the ministry who can help implement AI
- Third party experts can help implement AI
- AI will help minimize the use of paper [27]
- I think the public will benefit from introducing AI systems within the public service [47]
- I am innovative and ready for any change that might impact my role [48]
- I am aware that other ministries are using AI
- I am aware that other countries are using AI in their public service delivery [49]

7. **Benefits of AI include**

(choose as appropriate—strongly disagree/disagree/undecided/agree/strongly agree/I do not know)

- AI will increase collaboration with colleagues, external stakeholders and the public [50]
- AI provides data for informed decision making [51]
- AI is more reliable and consistent [52, 53]
- AI will eliminate repetitive tasks
- AI increases speed in working operations
- AI will enable multi-tasking and eases the workload for public workforce

- AI will help better understand patterns in the various public service sectors
- AI will assist the public 24/7
- AI will facilitate faster communication and response with the public

8. **Potential risk with the introduction of AI that will hinder AI progress** [54]

   (choose as appropriate—strongly disagree/disagree/undecided/agree/strongly agree/I do not know)

   - I am employed with the government; hence my job is secure and I do not feel threatened by AI technologies
   - I worry my job can be replaced by AI systems [55]
   - I am not informed enough about AI and AI adoption within my ministry
   - I believe the way data is currently collected by our ministry could be a barrier to the application of AI [56]
   - I believe the public will feel sceptical about using AI system
   - I believe the public needs more information about AI to trust new processes

9. **Possible changes that will take place with the introduction of AI** [57]

   (choose as appropriate—strongly disagree/disagree/undecided/agree/strongly agree/I do not know)

   - AI will change the current work processes
   - New jobs will be created within the Government
   - Public administrators will need to be retrained to upskill their current role [16]
   - Some jobs will be replaced by AI
   - AI will facilitate the roles of the public workforce [58]
   - AI will address more efficiently the needs of the public workforce in delivering quality public service

10. **The adoption of AI technologies within the ministries will be successful if:** [59]

    (choose as appropriate—strongly disagree/disagree/undecided/agree/strongly agree/I do not know)

    - The public workforce is informed and trained with the right skills in the use of AI
    - Increase awareness of AI among the public and the public workforce
    - The public is informed about the Government's AI initiatives and how these will improve the Maltese Public Service
    - AI is introduced as part of the Maltese education system early in order to have an AI knowledgeable future workforce
    - Preparation and planning for the introduction of AI is done strategically and for the long term
    - The implementation of AI systems is set against a robust legislative frame-

work to protect data privacy while promoting transparency and accountability
- The Government starts looking into the possible future economic risks related to AI adoption that might impact the Maltese population

References

1. CAHAI (2021) Artificial intelligence in public sector. Council of Europe, Strasbourg
2. Government of Malta (2014) Digital Malta 2014–2020. https://digitalmalta.org.mt/en/Documents/Digital%20Malta%202014%20-%202020.pdf. Government of Malta
3. Government of Malta (2014) Digital Malta: national digital strategy 2014–2020. Government of Malta, s.l.
4. Government of Malta (2019) MALTA the utimate AI launchpad; A strategy and vision for artifical intelligence in Malta 2030. Government of Malta, Valletta
5. Government of Malta (2019) Malta towards an AI strategy: high level policy document for public consultation. https://malta.ai/wp-content/uploads/2019/04/Draft_Policy_document_-_online_version.pdf. Accessed 12 June 2021
6. Government of Malta (2019) Malta AI. https://malta.ai/. Accessed 10 Aug 2021
7. Government of Malta (2019) Malta: the ultimate AI launchpad. https://malta.ai/wp-content/uploads/2019/11/Malta_The_Ultimate_AI_Launchpad_vFinal.pdf. Accessed 21 Aug 2021
8. Government of Malta (2019) Malta:towards trustworthy AI: Malta's ethical AI framework. MDIA, s.l.
9. Government of Malta (2019) Mapping tomorrow: a strategic plan for the digital transformation of the public administration 2019–2021. Office of the Principle Permanent Secretary (Office of the Prim Minister) and MITA, Valletta
10. Government of Malta (2019) Mapping tomorrow; A strategic plan for the digital transformation fo the public admnistration 2019–2021. Government of Malta, Valletta
11. Ayatollahi H, Bath PA, Goodacre S (2009) Paper-based versus computer-based records in the emergency department: staff preferences, expectations, and concerns. Health Inform J 15(3):199–211
12. Shekhar SS (2019) Artificial intelligence in automation. Artif Intell 3085(6):14–17
13. Dingli A, Haddod F, Kluver C (2021) Artificial intelligence in industry 4.0: a collection of innovative research case-studies that are reworking the way we look at industry 4.0 thanks to artificial intelligence.. In: Dingli A, Haddod F, Kluver C (eds) Studies in computational intelligence, vol 928. Springer Nature, Switzerland AG, pp 213–233
14. Pierce D, Shilling M, Forrest D (2015) Amplified intelligence. https://www2.deloitte.com/global/en/insights/focus/tech-trends/2015/tech-trends-2015-amplified-intelligence.html. Accessed 16 July 2021
15. Gibson M, Arnott D, Jagielska J (2004) Evaluating the intangible benefits of business intelligence: review & reserach agenda. s.l., s.n.

16. Wamba-Taguimdje SL, Wamba SF, Kamdjoug JR, Wanko CT (2020) Influence of artificial intelligence (AI) on firm performance: the business value of AI-based transformation projects. Bus Process Manage J 26(7):1893–1924
17. Veale M, Brass I (2019) Administration by algorithm? Public management meets public sector machine learning. In: Yeung K, Logde M (eds) Algorithmic regulation. Oxford University Press, Oxford, UK, pp 121–142
18. Bitkina OV et al (2020) Percieived trust in artificial intelligence technologies: a preliminary study. Hum Factors Ergonomics Manuf Serv Ind 30(4):282–290
19. Illanes P et al (2018) Retraining and reskilling workers in the age of automation. Mckinsey Global Institute, s.l.
20. Wright SA, Schultz AE (2018) The rising tide of artificial intelligence and business automation: developing an ethical framework. Bus Horiz 61(6):823–832
21. Krystal M (2021) COVID-19 the upskilling imperative: building a future-ready workforce for the AI age. Deloitte, Canada
22. Mikhaylov SJ, Esteve M, Campion A (2018) Artificial intelligence for thepublic sector: opportunitiesand challenges of cross-sectorcollaboration. Philos Trans R Soc A 376(20170357, 27 06):1–21
23. Criado JI, Gil-Garcia JR (2019) Creating public value through smart technologies and strategies: from digital services to artificial intelligence and beyond. Int J Public Sect Manag 32(5):438–450
24. Van Buren E, Chew B, Eggers WD (2020) AI readiness for government: are you ready for AI?. https://www2.deloitte.com/global/en/insights/industry/public-sector/ai-readiness-in-government.html. Accessed 16 July 2021
25. Capita (2019) Future of work: robot wars or automation alliances? Capita, s.l.
26. Tomar L, Guicheney W, Kyarisiima H, Zimani T (2016) Big data in the public sector. Inter-American Development Bank, s.l.
27. Jöhnk J, Weißert M, Wyrtki K (2021) Ready or not, AI comes—an interview study of organizational AI readiness factors. Bus Inf Syst Eng 63(1):5–20
28. Chwelos P, Benbasat I, Dexter AS (2001) Reserach report: empiricial test of an EDI adoption model. Inf Syst Resour 12:304–321
29. Bertrand A (2020) Why AI and the public sector are a winning formula. https://www.ey.com/en_gl/government-public-sector/why-ai-and-the-public-sector-are-a-winning-formula. Accessed 21 June 2021
30. Bertrand A (2020) Why AI and the public sector are a winning formula. https://www.ey.com/en_it/government-public-sector/why-ai-and-the-public-sector-are-a-winning-formula. Accessed 30 June 2021
31. Aboelmaged MG (2014) Predicting e-readiness at firm-level: an analysis of technological, organizational and environmental (TOE) effects on e-maintenance readiness in manufacturing firms. Int J Inf Manage 34:639–651
32. Yum JJ, Zhao X, Jung K, Yigitcanlar T (2020) The culture for open innovation dynamics. In: Sustainability in 2nd IT revolution with dynamic open innovation, vol 12, no 12, p 5076

33. Eggers WD, Agarwal S, Kelkar M (2019) Government executives on AI. https://www2.deloitte.com/global/en/insights/industry/public-sector/ai-early-adopters-public-sector.html. Accessed 12 July 2021
34. Esteve M, Campion A, Gasco M, Mikhaylov S (2020) The challanges and organisational factors in collaborative artificial intelligence projects. Soc Sci Comput Rev
35. Valle-Cruz D, Sandoval-Almazan R (2018) Towards an understanding of artificial intelligence in government. Association for Computing Machinery, Delft, The Netherlands
36. Valle-Cruz D, Ruvalcaba-Gomez AE, Sandoval-Almazan R, Criado I (2019) A review od artificial intelligence in government and its potential from a public policy perspective. In: Proceedings of the 20th annual international conference on digital government reserach, pp 91–99
37. Valle-Cruz D, Ruvalcaba-Gomez AE, Sandoval-Almazan R, Criado IJ (2019) A review of artificial intelligence in government and its potential from a public policy perspective. Association for Computing Machinery, Dubai, UAE
38. Pencheva I, Esteve M, Mikhaylov SJ (2020) Big data and AI—A transformational shoft for government: so what next for research? Public Policy Adm 35(1):24–44
39. Risse M (2019) Human rights and artificial intelligence: an urgently needed agenda. Hum Rights Q 41(1):1–16
40. Sun TQ, Medaglia R (2019) Mapping the challenges of artificial intelligence in the public sector: evidence from the public healthcare. Gov Inf Q 36(2):368–383
41. Bughin J (2018) Marrying artificial intelligence and the sustainable development goals: the global economic impact of AI. https://www.mckinsey.com/mgi/overview/in-the-news/marrying-artificial-intelligence-and-the-sustainable. Accessed 19 July 2021
42. Manyika J, Bughin J (2019) The coming of AI spring. https://www.mckinsey.com/mgi/overview/in-the-news/the-coming-of-ai-spring. Accessed 19 July 2021
43. Harrison-Prentice T et al (2021) Navigating the path of AI adoption: capturing busines value by identifying critical success factors for AI adoption. https://www2.deloitte.com/content/dam/Deloitte/nl/Documents/strategy-analytics-and-ma/deloitte-nl-sa-and-ma-navigating-the-path-for-ai-adoption.pdf. Accessed 19 July 2021

Appendix D
Step-by-Step Breakdown of the Research Process for Auditability Purposes

1. Malta: The Ultimate AI Launchpad—The Malta AI Strategy launch
2. Lack of information about the activities identified in the strategy
3. Research question: What is the extent of AI readiness in the Maltese public administration?
4. Research literature review re AI readiness and practices by foreign governments
5. Due to the lack on local information, decided research study will apply a mixed method with the aim to gain insight from the experts implementing AI in the public administration and understand the perception of AI from their workforce
6. Decision to interview all the 14 CIO engaged with the ministries. Some CIOs have more than one ministry under their responsibility
7. Decision to interview 4 ICT experts from the public sector. The experts chosen are assisting the government to implement AI nationwide
8. Inform OPM about my research to get access to the target population. The OPM DPO was provided with a copy of the research proposal and the approved ethical and consideration form from IDEA Academy
9. Design interview questions from the literature review
10. DPO approved participation. Provided a copy of the consent form and the interview questions addressed to the experts for approval
11. DPO informed the experts about the research and disseminated the consent form and the interview questions
12. The experts contacted me via email to set interview dates
13. 18 interviews conducted over 2 months
14. Transcription details were sent to the participants for approval of text
15. Process of data analysis: inductive thematic analysis by extracting the maximum excerpts, sort in collective themes, identify sub-themes, themes and overarching themes
16. Inform DPO about the survey addressed to the public administration and provide the opening consent paragraph
17. Design the survey questions from the quantitative analysis

M. Sciberras and A. Dingli, *Investigating AI Readiness in the Maltese Public Administration*, Lecture Notes in Networks and Systems 568,
https://doi.org/10.1007/978-3-031-19900-4

18. Provide DPO a copy of the survey questions for approval
19. DPO disseminates the link to the survey question via the ministries, who were instructed to share with all the public service and public sector employees
20. The survey was in circulation for 6 weeks
21. Process of data analysis: evaluate the information received and present in a graphical manner
22. Mixed-method data analysis: validating qualitative data via triangulation design. Overarching themes identified are evaluated vis-à-vis the survey replies, the National Strategy and literature review. Excerpts for validation and reliability purposes are presented
23. Propose a set of recommendations based on the outcome of the triangulation analysis.

Appendix E
Excerpts for Overarching Theme 1: Understanding AI

Overarching theme 1: Understanding AI		
Theme	Subtheme	Excerpts
1.1 AI as a Tool	1.1.1 Powered by data	05: "… AI cannot exist or function properly, effectively without data, and it needs lots of data, good quality data and real time data."
		08: "… we have a good potential set of data set to use for AI."
	1.1.2 Assistive to humans	02: "… it is a tool that we can use as public employees and for the professionals working IT, in order to further enhance the business needs within the respective policy area."
		08: "… meant to facilitate the way we do things, improve the way we do things and help us make our life better."
		18: "AI should be design to be assistive to humans."
		06: "… AI, it will make life easier on the front end. Like for instance, the manager, it needs to take a decision or she needs to take a decision and the AI will assist …"
		14: "AI is a tool that will help and accommodate customers, clients, public administration, to enhance business services."

(continued)

M. Sciberras and A. Dingli, *Investigating AI Readiness in the Maltese Public Administration*, Lecture Notes in Networks and Systems 568,
https://doi.org/10.1007/978-3-031-19900-4

(continued)

Overarching theme 1: Understanding AI		
Theme	Subtheme	Excerpts
		13: "… if done properly should not create a disaster but rather eliminate mundane repetitive tasks and facilitate simplification through the adoption of mechanical processes …"
	1.1.3 Autonomous	03: "… humans we need to learn what is the AI concept in relation to automation of specific tasks and another aspect is in relation to intuition and machine intelligence which is derived by the human input."
		09: "A set of new rules in advanced software development, that to a larger extant, are able to execute multiple functionalities."
		06: "… software algorithms to automate things …"
	1.1.4 Intelligent	13: "… automation of processes in a more intelligent way …"
		11: "… simulation of intelligence in machines that are programs to learn …"
		16: ".. system that is able to carry out certain human cognitive abilities …"
		04: "… it is a means of facilitating anything which cannot be processed by human power but if processed with AI it can be tested from a level of institutional knowledge."
		10: "… collection of technologies that simulate the human intelligence in machines …"
		07: "… just making it possible for machines to assist us in our daily lives and be able to go see information that otherwise humans cannot"
1.2 Current uses of AI	1.2.1 Cloud	13: "Complete cloud printing solution …"
		16: "From an infrastructure perspective, especially when moving towards cloud solutions, I think we're at a good level, so if certain data manipulation is to be carried out, there are the services that we need."

(continued)

(continued)

Overarching theme 1: Understanding AI		
Theme	Subtheme	Excerpts
	1.2.2 Big Data	02: "… for the AI project, the big data project that we are developing, we are going to use a power BI, …"
		11: "… business portal …"
		16: "Business intelligence (BI) systems are also crucial because when you have a data warehouse, basically you're building your history."
		16: "The problem is that a lot of entities and ministries don't even have BI in place."
		13: "… major pilot project across the entire government for remote working"
		13: "… enhance metering system which is intelligent enough to inform the customer …"
		14: "… services are being upgraded with various solutions, which incorporate AI."
		04: "… the ministry has numerous systems in place targeting project management, resource allocation or resource management."
		16: "The data warehouse can be applied to build intelligence, using the data, by crossing into different methods, joining data together for reporting purposes, visualization purposes, there are a lot of things that you can do with it."
		05: "IOT devices are that fundamental component, that they capture real data in real time and in big quantity."
	1.2.3 Semantic analysis	15: "Another big area, which I am seeing a lot of is the use of a text pattern recognition in terms of sentiment analysis and other similar users. We get feedback from the citizen on what type of policies are needed or feedback on policies, and feedback is received from posts and the social media discussions that are going on, that you can get further information …"
		07: "… that all the laws will be in a machine readable format."

(continued)

(continued)

Overarching theme 1: Understanding AI		
Theme	Subtheme	Excerpts
		08: "… sentiment analysis to identify other people's opinions about certain matters …"
	1.2.4 Chatbots and NLP	08: "… portal for the agency is using AI basically based on questions, asked and information gathered and machine learning techniques from the general public."
		16: "… to build and language technology platform for Maltese and English who would use this as a machine translation."
		11: "… chat bots with AI power to answer queries from the portal clients …"
		12: "… we are implementing chatbots …"
	1.2.5 Different AI applications	04: "… NFC enabled cards, …"
		05: "… data is collected through IOT props … We also use satellite data, high resolution, satellite imagery, which are taking periodically, and they gave us information … he opens the camera with augmented reality, the camera will tell him exactly where to stop, take a picture and send it to us …"
		08: "… a huge dataset of photos from the department of information …"
		10: "… robotics—specialized machines that assist …"
		13: "… completely automated transportation system."
		11: "… assisted search through AI, to assist users, to easily find the documents he or she searching …"
		12: "Currently, we have one parking spot which has sensors on the floor, there are signs reporting the availability of parking… irrigation systems in public gardens with in-self probes and the soil would calculate the amount of humidity and other values … through IoT and AI we know if the bin is full or not …"

Appendix F
Excerpts for Overarching Theme 2: Policy and Procedures

Overarching theme 2: policy and procedures		
Theme	Subtheme	Excerpts
2.1 National AI strategy	2.1.1 Pros and Cons	17: "... what Malta is achieving in regards to implementing AI is great."
		18: "The strategy is 10yrs long and the document is clear on what the deliverables will be, however it is missing a concrete implementation plan."
		01: "... as non-private employer, the public service has responsibilities that will have social implications and obligations. However, if we look into the immediate future, there is no overarching vision, there is no strategic approach, and that can actually be misguided by the market ..."
2.2 Ministry's own	2.2.1 Strategic vision	10: "... strategy of the ministry is based on the Malta's National AI Strategy ..."
		01: "does have an IT strategy, which does foresee the implementation of modern-day technologies but it does not specifically discuss AI technologies and their implementation."

(continued)

M. Sciberras and A. Dingli, *Investigating AI Readiness in the Maltese Public Administration*, Lecture Notes in Networks and Systems 568,
https://doi.org/10.1007/978-3-031-19900-4

(continued)

Overarching theme 2: policy and procedures		
Theme	Subtheme	Excerpts
		16: "Ministerial strategies shouldn't be longer than two to three years. Goals can be adjusted accordingly and monitored for the expected results. Ministries having their own strategies is encouraging as it shows the drive of the respective ministries towards successfully achieving these AI initiatives."
		04: "I, as a CIO, have to have a strategy. I drew a strategic document, which is tagged with the level of organisational chart that I wish to have, with the level of work of each level, of how it is going to affect my unit. Basically, the more systems you have, the more management and policies you have to adapt."
		14: "… technology nowadays … must be defined within the process of aligning services too."
	Staying focused	17: "… we should specifically identify a niche within AI. I think the strategy did this strictly, it wants to set up Malta as the testbed, inviting operators to come to Malta, develop and test their innovation here."
		14: "… are implementing new solutions with the aim to have them operational across other ministries and integrate or replace the existing older systems in order to have a standard transition."
		07: "… do the change in the design and the digital architecture, obviously where AI could be of benefit is being identified to a DBR exercise. Eventually they will become more tangible."

(continued)

(continued)

Overarching theme 2: policy and procedures		
Theme	Subtheme	Excerpts
	2.2.3 No ministerial strategy	03: "Currently the ministries have no strategy in place but there is a vision, a direction we want to take. … We are starting from the Business Intelligence to have all the data in one place, then we can move forward from there."
		09: "Currently, there is no strategy in place. A strategy will be drawn up once the processes are being implemented."
		06: "… how things work in the public service most of the times is that, there's not like a holistic view of what is going to happen."
		06: "… there will be issues with the workforce, especially during the transition. However, I'm not aware that at least within our ministry, we are already looking at what will happen to the workforce."
	2.2.4 In house strategy—Evidence based	05: "… officers at the ministry with highlights, insight and suggestions to help taking the best decision at the best time. The advantage of having AI in our ministry is that it has the tech to detect problems or possible problems at a very early stage …"
		08: "… will give us more data driven decisions."

Appendix G
Excerpts for Overarching Theme 3: Data

Overarching theme 3: data		
Theme	Subtheme	Excerpts
3.1 Data availability	3.1.1 Data collection	14: "…data gathering and data mining is the essential part to assist public and clients all over the ministry."
		13: "Most of our systems are connected to a server, a database that is commonly used, of which structures are according to big data requirements. In this case, field names are synchronised easily."
		01: "The manually collected information is then transposed digitally into the system by a human. The data is not native digital in this case. There is a lot of data input or manual errors due to the multitude of different formats, multitude of different documents that you're dealing with each application,"
		05: "Any process which was paper-based, now it is fully digital."
		09: "Most of the data is collected by paperwork and input in Excel sheets."
		13: "Completely registry and digitalised filing solution; basically, we are a completely paperless office."
		08: "… that were collected digitally, but that does not mean that the data is ready to be used for AI yet for an AI project …"
		12: "… So yes, everything is digital."
		14: "… data is collected through third parties, namely network operators. This information is digitally available."

(continued)

M. Sciberras and A. Dingli, *Investigating AI Readiness in the Maltese Public Administration*, Lecture Notes in Networks and Systems 568,
https://doi.org/10.1007/978-3-031-19900-4

(continued)

Overarching theme 3: data		
Theme	Subtheme	Excerpts
		16: "The good thing is that we are moving to digital solutions. That's quite positive. And I think during the last few years there has been quite an improvement in that regard."
		06: "Most of the information is collected through these forms"
		11: "… data collection is done through various databases."
		14: "… information is at hand in minimal timeframes, we are talking about seconds. This data is retrieved from the databases of all the Maltese network operators."
		06: "… data … is relatively small and does make it difficult to apply AI to it."
	3.1.2 Data structure	01: "… we do not have the correct data structures, we do not have the correct data quality—data quality is as in availability, as in semantic correctness, as in error free data."
		08: "We need to have a baseline …"
		15: "… collecting, structuring and making it available, obviously where legally possible when feasible, but it's important."
		07: "We want to be agile, and unfortunately the current procurement system and methodology, does not permit us to be agile enough to implement."
		08: "… amount of data that needs to be to go into the systems is massive …"
		13: "Technically speaking, I have not found one way, or a solution that does not require reformatting once it is needed to actually do something."
		15: "… the other part would be that the data exists, but it has to be cleaned, it has to be sanitized, it has to be formatted. So, with some cleaning, it can be done. In those areas, it's much easier to trigger a project."
		12: "… first collected data, structured data, not just data, what we call structured data so that when we can connect and interconnect systems …"
		08: "It's too complex because there is a lot of unstructured data …"
	3.1.3 Data dumping	18: "very rich in generating volumes of data, however, this data is being dumped as it is humanly impossible to analyse, hence the need for AI processes to reap knowledge"
		12: "Although we collect structured data, we don't make hundred percent use of this structure data, but the data is there."

(continued)

(continued)

Overarching theme 3: data		
Theme	Subtheme	Excerpts
	3.1.4 Data quality	04: "… is the information gathering process and the information itself sustainable? Information is rarely assessed in this manner."
		15: "… the other part would be that the data exists, but it has to be cleaned, it has to be sanitized, it has to be formatted. So, with some cleaning, it can be done. In those areas, it's much easier to trigger a project."
		13: "It is a crucial step to have AI in place with a uniform database, uniform and standard infrastructure for data collection for increased collaboration."
		02: "The data needs to be formatted into something which can be AI and machine readable. Another important thing is that we need to collect information from those same interactions that the client is having with AI for the model to learn. In relation to the MTA project, some of the data available is partially digital."
		08: "… our data and the quality of our data is a huge challenge …"
		14: "important to administer a garbage-in garbage-out concept. In our department, data cleaning is done regularly."
		11: "Data sanitisation from these processes is most important before starting to implement AI."
		11: "The government has a wealth of data. That's why I've been emphasizing the importance of data sanitisation and data cleansing …"
		14: "If the wrong data is fed to the system, then the output will be incorrect. Data entry needs to be accurate and correct."
		13: "A nationwide data cleanse to allow ministries to have a single common platform to be able to integrate …"
		08: "… apart from more concise and better decision-making and more efficient decision-making if obviously the data is clean and everything depends on the quality of the data …"
		08: "… personalised services, the federated learning, they will come, they will be there, but we need to have clean data, and this service is going to be our biggest challenge, how we're going to connect all these systems, because we're going to find a lot of people who are going to challenge us legally …"
		08: "… more focused on getting our data set, ready to start embracing AI."
3.2 Data Processing	3.2.1 Legal framework	14: "… need data available across the board. Legacy systems currently make it impossible to operate in such manner …"

(continued)

Appendix H
Excerpts for Overarching Theme 4: Preparedness

Overarching theme 4: preparedness		
Theme	Subtheme	Excerpts
4.1 Ready or not	4.1.1 Not ready	13: "Yes and no. It needs a culture change."
		01: "The ministry is not ready, it is decades from being ready. There's the wrong mentality, it is the wrong culture."
		09: "The ministry is not as yet ready and we are finding some resistance, however we are trying to get everyone on board."
		08: "They are not yet ready for sure, they've heard about these solutions and about these technologies, but they are not yet ready nor their mindset …"
		06: "I don't think that the ministry is ready. Let's say it has the like around 30%, …"
		13: "This is not a magic wand situation, as it is very complicated and the governments' system is not ready for AI."
	4.1.2 Getting there	10: "… ministry is on the right track to be ready."
		07: "… it depends on the department that you are working in."
		11: "… within my ministry, we adapt to change. I mean, I'm sure there will be some resistance …"
		16: "There is a lot of work to be done still but we are moving in the right direction."
	4.1.3 Ready to implement	05: "I think we were one of the first ministry to be ready to implement earlier … … We were the first ministry to use augmented reality and sort of AI in the government. So, let's say the pioneer of the government."

(continued)

M. Sciberras and A. Dingli, *Investigating AI Readiness in the Maltese Public Administration*, Lecture Notes in Networks and Systems 568,
https://doi.org/10.1007/978-3-031-19900-4

(continued)

Overarching theme 4: preparedness		
Theme	Subtheme	Excerpts
		02: "There are always the early adopters, champions I call them and those who prefer to wait, which is normal. So, you normally have a mixture of different people. Initially we focus on the early adopters, but those are the people who will sell your product. If you enjoy the experience for sure, you're going to talk about it. So other people will start using it. It's a whole process. The challenge with technology is always changing the culture."
		15: "... it will take time to affect the actual changes because transformation doesn't happen just by implementing a system. It takes mentality change. It takes training, it takes changing to the roles and responsibilities."
4.2 Technicalities		16: "The technological aspect is the level we are most advanced in as the infrastructure is already in place."
		15: "Technically I believe we are. We have the capabilities of pushing the technology as much as the business owner wants from the technology readiness..."
		14: "I can say no. In fact, to adopt such systems solutions, I had to go for a third-party service. We had to deploy the services. We are still deploying third party services because the technicalities for such solutions cannot be taken onboard by my staff definitely."
		16: "... preparation for the influx of upskilling that will be needed by the ministries."
		16: "From a technical perspective, we're upscaling our people to get acquainted with several programming languages associated with data science."
		07: "We want to be agile, and unfortunately the current procurement system and methodology, does not permit us to be agile enough to implement."
4.3 Availability of funds		10: "... the more complex projects would need a more detailed financial analysis and possibly they would spread across a number of areas and a substantial investment would be required."
		13: "Yes. This on a case-by-case basis"
		14: "For budgeting reasons, approximate costs for systems' implementation are identified a year earlier."
		16: "Financial wise, sometimes it is difficult to define the exact cost of a solution, however, that is where certain expertise is involved to provide an estimation on the cost of systems."
		10: "... investment needed to have the necessary architecture in place..."

(continued)

(continued)

Overarching theme 4: preparedness		
Theme	Subtheme	Excerpts
4.4 Competitivity		04: "In terms of management, in terms of employee satisfaction and in terms of how is the organisation and the employee benefiting realistically, we are losing a number of employees within the public service because of we cannot compete with the private sector or even with the entities as the government salary scale is lower than the market competition."
		13: "This is a very fluid, highly paced and highly remunerative environment that we cannot keep up with unless you specifically invest in that resource environment that is usually ideal for the private sector business model not the public sector."

Appendix I
Excerpts for Overarching Theme 5: Change What?

Overarching theme 5: change what?		
Theme	Subtheme	Excerpts
5.1 Current Practices	5.1.1 Set in old ways	14: "… there was some resistance because it was a culture shock for each individual. When people are customary to paper-based set ups, it proved difficult to change their attitude towards technology."
		08: "We still have to go through quite a lot of culture change and training and giving them more awareness of how this can make their life at the place of work better."
		13: "Culture-wise and mentally, we are 10 years behind."
		02: "… cultural adaptation for me is key."
		08: "It depends mainly on having the right resources, ready, not to adapt themselves to the culture, change to habits and finding resources is becoming a challenge."
		01: "We will not be able to transform vertically or horizontally into a digitally driven or AI driven public service, unless the culture changes and culture is difficult to change."
		14: "… the team in my remit is. However, it needed a culture change."
		09: "… to implement a digital transformation but first it needs a culture change."
		06: "… with the culture and the mentality needs to change a lot, probably that will be the most difficult part."
		13: "Yes and no. It needs a culture change."

(continued)

M. Sciberras and A. Dingli, *Investigating AI Readiness in the Maltese Public Administration*, Lecture Notes in Networks and Systems 568,
https://doi.org/10.1007/978-3-031-19900-4

(continued)

Overarching theme 5: change what?		
Theme	Subtheme	Excerpts
		08: "… the preparation it's in our culture now. … … When I am introducing a change, I have to involve the people who are going to be affected by the change from day one."
		01: "There's this public service perspective that dislike change < as in, but we've always done it this way > . I have an army of diplomats and diplomats are people who love to work with pen and paper."
		17: "Hindrance is not from a technological point of view, more from a social-political point of view."
		06: "… these people would be when implementing change, that they will resist change. Very often they will be very negative …"
		16: "AI is very vast and I think it's the culture the problem."
		15: "… the biggest challenge would be the change management involved after implementation …"
		10: "… even change how existing public services are currently being delivered …"
		16: "Having people with basic skills in the area will help transform processes but people are not easy to adjust to change …"
		07: "We want to be agile, and unfortunately the current procurement system and methodology, does not permit us to be agile enough to implement."
	5.1.2 Institutional structure	05: "… there have been a new position, like a new director and assistant director, which role is mainly to work on AI and technology project with the collaboration, with my office."
		07: " One can't expect that someone who takes care of operations will also take care of AI, even at expertise level, as obviously you need a certain level of expertise in AI to be able to make full use and benefits of what AI could give you."
		08: "… problems that we have is we have a lot of managers but we don't have soldiers, foot soldiers. That's a problem because a lot of these people who we are paying at in scale five at senior management level … are still doing operational work rather than leading innovation …"

(continued)

(continued)

Overarching theme 5: change what?		
Theme	Subtheme	Excerpts
		13: "I do not believe we require to specifically employ involve these experts permanently within our workforce."
		16: "I find it very positive that we have a strategy, and we will have a new public administration strategy, I still think that this extra level is still missing from both the horizontal level and administrative level
	5.1.3 Diffuse misconceptions	13: "AI is not like it is presented in movies, this wrong perception is causes problems as people are not technologically informed and are either over sensitive, under sensitive or have the wrong expectations altogether."
		16: "There's a lot of misconception that with AI, that will be a loss of jobs. I'm not saying that in certain sectors, it's not the case because I would be unrealistic, and in the public administration, that's not the aim."
		02: "AI will not completely take over the human …"
		10: "… delusion that AI and machines will one day take over the human race, when in reality, the sole purpose for AI is to support the human race by giving solutions to complex problems that it would take a number of years for the human mind to accomplish."
		14: "Even though AI is playing an important part, the human side is very predominant …"
	5.1.4 Increase awareness	07: "… any kind of method that you introduce, you need to get trust, trust from the users to use them."
		01: "If implemented correctly, the administrative staff, the client, wouldn't know there's AI behind the system. … but at the clerk level or user level, for them it doesn't actually matter if it is AI or a human behind the screen processing information."
		10: "… we need to start reaching out and educate our workforce more on AI and it's potential."
		13: "I believe the first approach would be teaching, informing and then specifying the benefits of such systems."
		17: "… awareness but awareness among the different project owners."
		08: "We're looking at in novel ways, how we market the public service …"

(continued)

(continued)

Overarching theme 5: change what?		
Theme	Subtheme	Excerpts
		09: "… information is key to AI acceptance among the public workforce and the public. I believe the public workforce and the public need assistance to develop …"
		15: "… one of the human landscapes is this awareness and these letting things settle down that, listen, you will not be fired. We are pushing, right, because it gives a benefit so that the human part is important."
		06: "… that's a lot of training, a lot of awareness will have to be done. I'm sure that don't matter how much awareness you try to do to instil in our employees, it's not going to be taken up 100%, …"
	5.1.5 Dissolve silos	16: "… work horizontally with the different ministries, and the team can learn from different projects in different ministries."
		15: "Data sharing would eliminate the silos. It's too structured in the way ministries function. So, the possibility to data share will actually remove working in silos."
		07: "… obviously we are looking about how our systems connect in a more holistic way rather than having systems in silos and obviously to that process a full DBR exercise of the different processes is envisaged to happen."
		08: "No federated learning coming out from what we are looking at, creating a service bus, where finally we start breaking the silos between ministries and we'll have systems which are interconnected to the service bus, sending data to the citizen …"
		13: "A shared solution across all ministries that will allow for data crunching, consolidation, digestion and sharing."
		11: "We shouldn't work in silos anymore, this is one of the most disadvantages that the government, as yet, we still work in silos; although a lot of improvements were done in the last years …"
		03: "… share data as one and eliminate the working in silos approach"
		11: "… silo mentality, which needs to be eliminated and the one government concept …"

(continued)

(continued)

Overarching theme 5: change what?		
Theme	Subtheme	Excerpts
		11: "… seeing the importance of the have to work together, the one government concept, which I did with the project, whereas entities four needs to collaborate together to achieve this massive project …"
		11: "Data sharing is going to be a very difficult task to do due to our traditional methods of working …"
5.2 To drive AI	5.2.1 Positivity prerequisite	08: "… AI could, if it's sold positively, leave a positive impact …"
		18: "… clearly show the community the benefits that such technology will bring about and the positive impact it will leave in people's lives."
		14: "Technology is embedded into our lives and AI can be looked at as an assistive technology facilitating our needs."
		04: "The ministry is ready. We are using positive concepts in terms of changing culture. We inform our workforce in terms that going digital is not changing the current processes, but facilitating their workflow."
		13: "It is a change process which we need to encourage through open and clear communication and for a positive and humane approach."
	5.2.2 Leadership	10: "Get Business owners onboard."
		02: "We have a lot of key players and leaders within various areas that are leading change and leading innovation."
		05: "… when he became a permanent secretary of the same ministry, it really helped this ministry to move forwards faster because obviously he brought his vision where it mattered."
		13: "Analyse-Document-Simplify is the way I implement change."
		02: "There are always the early adopters, champions I call them and those who prefer to wait, which is normal. So, you normally have a mixture of different people. Initially we focus on the early adopters, but those are the people who will sell your product. If you enjoy the experience for sure, you're going to talk about it. So other people will start using it. It's a whole process. The challenge with technology is always changing the culture."

(continued)

(continued)

Overarching theme 5: change what?		
Theme	Subtheme	Excerpts
		13: "… get the stakeholders and decision makers at the top levels to have expectations that make sense and the people on the other end of the spectrum to understand what it is and how such systems can help them out, without creating a dogma about it."
	5.2.3 Public inclusion	08: "They know I went to hospital, they know that I worked, I was at work. They know I got injured at work. So they are prompting me to apply for in general benefit. So do I have a big brother here? So these are the challenges."
		18: "to truly be successful, we need to make sure that the public is understanding and reaping the benefits out of these pilot projects."
		14: "… the public does not know that AI is assisting in their requirements. However, the public should be educated about AI as a concept and relate it to their everyday activities."
		13: "When it comes to AI, the biggest issue is ignorance in the subject matter. We are not informing the public that it's happening or how it's happening."
		06: "If people are not going to be prepared for this big shift, they are going to be a lot who will suffer in some way or another financially or like they will be left out like out of society, that they will become marginalised."

Appendix J
Excerpts for Overarching Theme 6: In Demand

Overarching theme 6: in demand		
Theme	Subtheme	Excerpts
6.1 Knowledge and Expertise	6.1.1 Lack of it	03: "The government lacks a central repository of knowledge"
		14: "… don't have the technical knowledge to perform such things."
		16: "One very important element is the up-skilling aspect, as lack of technical knowledge limits achievement in AI adoption."
		13: "Key stakeholders need to be more informed as these currently do not have a clear picture on what to expect and the proper milestones to look out for …"
		16: "I think there still is a lot to be done to reach a certain level. I think the problem falls with the lack of knowledge within the respective ministries. It's very difficult to transform something from a management perspective, from an operation perspective, from a technology perspective, if you don't have people who have worked in this domain or have studied in this domain, who can direct you."

(continued)

M. Sciberras and A. Dingli, *Investigating AI Readiness in the Maltese Public Administration*, Lecture Notes in Networks and Systems 568,
https://doi.org/10.1007/978-3-031-19900-4

(continued)

Overarching theme 6: in demand		
Theme	Subtheme	Excerpts
		06: "… it always boils down to the lack of personnel and the lack of having knowledge, human resources."
		08: "The junior software engineer will come to the government, spend a couple of years getting experience and will leave. So you're not building foundations there because they will find out that he is being paid peanuts."
		14: "I don't think people understand that concept, that AI will assist them and their operations."
		16: "… specialisations require specific attributes and proficiencies of which the public service lacks."
		10: "… there is a lack of expertise and the lack of knowledge …"
		16: "… I think we need to start by having knowledgeable individuals on board, and knowledge is not easy to get."
		16: "… the ministries do not lack the knowledge but more varied competencies are needed."
		04: "… if there is a change in a headship position, or a change in key individuals, there is no knowledge transfer and we lose a level of institutional knowledge."
		16: "… trying to transform something, I think it's important that the people on board at least have basic competency in the area that you are focusing on."

(continued)

(continued)

Overarching theme 6: in demand		
Theme	Subtheme	Excerpts
		07: "… there is already the IPS which does a lot of courses, but for management, I would make it mandatory for management to attend to an AI specific course for awareness, and I would demand that at least they identify one AI project that they could plan in the long term."
		11: "Training should not be done only from the IT aspect but also from the business owners point."
		13: "AI is a learning process, it is all about skills and teaching people."
		17: "… there's a lack of education in general, amongst the public service—though knowledge is now increasing …"
		07: "… create a specific webpage/presence, whatever, for the public service that promotes AI specifically. I would promote certain AI initiatives that could be used. There could be even launched as open source. It could showcase identified AI approaches within the government, …"
		06: "… many people who are working in government, they just believe that everything will go smoothly. I am sure that will not be the case."

(continued)

(continued)

Overarching theme 6: in demand		
Theme	Subtheme	Excerpts
	6.1.2 ICT class	06: "… to then implement the actual algorithms, to write them, to implement them, to see, to fine tune them, that will require a lot of people who are very knowledgeable. It's not just a matter about IT people, we are not talking about a number of IT people and shifting them to these new roles, it doesn't work like that. The work will be much more difficult."
		16: "… ensure that we don't lose the skillset that we have. I think this is very important even in the private sector."
		03: "People move, especially within the public service."
		07: "… the ICT Class consists of 8 standards roles with job descriptions for IT and common eligibility requirements which IMUs within Ministries use to recruited IT People. This is somehow strict for example there is a Systems Administrator identified in the class so must stick to that, we cannot for instance issues a call for let's say an AI Developer."
		13: "What we mostly need from our end is ICT departments with people that are knowledgeable enough to understand processes so that we can understand the potential of the creating of new projects and then to be knowledgeable enough to be able to manage it to completion."

(continued)

(continued)

Overarching theme 6: in demand		
Theme	Subtheme	Excerpts
		06: "… in theory we have the structure, but then we have so many people moving from one ministry to another, they're leaving for the private sector that in most cases, lots of people started doing a bit of everything."
		08: "… have about 150 people in the ICT class at the moment in the public service, there are seven, eight grades starting from scale 13 to scale five, like the managers, you know, and one of the problems that we have is that we have a lot of managers but we don't have soldiers, foot soldiers …"
		13: "… we have invested in our ICT so that we are able to create a healthy mix of projects that are both developed internally and others that we request the aid of the private sector."
		14: "The ICT class needs more human resources, with specific expertise to meet the needs of the ever increasing demand within the ministry."
		14: "The course that the government is pinpointing to staff within the ICT class, those are very relevant. The ICT class within the government is very relevant to the operations of the government. And obviously coming from me, I think the ICT within the government needs to be done because we are increasingly reliant on technology."
		02: "… we need to take the next leap in specializing."

(continued)

(continued)

Overarching theme 6: in demand		
Theme	Subtheme	Excerpts
	6.1.3 Role of the CIO	07: "… from the IT side, we identify opportunities ourselves. Unfortunately, the focus currently is through my office. We identify certain areas where within the data that could benefit from it. It's not the other way around where the business identifies the needs and comes to us. Ideally it's like that."
		14: "The IMU team assist business owners to implement solutions, but we are not the business owners ourselves. This is the difference that needs to be clearly communicated among the public workforce."
		07: "… not the other way around where the business identifies the needs and comes to us. Ideally it's like that. But unfortunately we still need to push our workforce to understand the benefits that could be achieved …"
		14: "Most of the IMU team comes from the technical side but the administrative side took over our role."
		14: "… the ministry does not have the manpower to cater for such services in their niche. It's practically impossible …"
	6.1.4 Outsourcing: collaboration or dependency	09: "As an IMU we do not have the expertise, usually such expertise is outsourced. We do not have to train staff in AI and BI."
		08: "… because we don't have resources, we outsource …"

(continued)

(continued)

Overarching theme 6: in demand		
Theme	Subtheme	Excerpts
		13: "… engage the private sector when dealing with particularly complicated solutions as this ensures the enrolment of the best key experts we can avail from … … When implementing a project, usually we employ experts and the experts would be data analysts and risk assessment people, …"
		06: "… we will define the specifications, the requirements, but then we'll be outsourcing them."
		14: "I can say no. In fact, to adopt such systems solutions, I had to go for a third-party service. We had to deploy the services. We are still deploying third party services because the technicalities for such solutions cannot be taken onboard by my staff definitely."
		08: "..engage a third party assistance in terms of data engineering …"
		16: "Innovative public and private administration—two folds—public projects can be handled by a private subcontracting—a hybrid approach."
6.2 Educational movement	6.2.1 Job retention	01: "… AI knowledge will not replace people who don't have the AI knowledge, but people that have an AI knowledge will actually be the first runners to making cues and making effective use of that technology."
		11: "… we train our workforce for AI. We train our users to use the AI as well. And we can introduce our users for AI. This would help for the overall strategy, not just for the government …"

(continued)

(continued)

Overarching theme 6: in demand		
Theme	Subtheme	Excerpts
		06: "... I find it a difficult, for example, to people over 50 years old to be re-skilled so easily."
		06: "... but not just manual work will be at risk."
		10: "... be some job displacements for jobs which are repetitive in nature and which ultimately can be automated ..."
		13: "Knowledge is essential for our workforce to efficiently make use of AI."
	6.2.2 Career paths	16: "... we need to be financially competitive to do that. It's a career path that should be exploited for the country's benefit."
		10: "AI can also create new jobs."
		13: "New job creation with the AI as the new expected industrial revolution."
		10: "Encourage courses and scholarships to employees in the ICT Class."
		11: "Another aspect which needs to improve as a government as a whole is human resources in specialised fields, or training must be provided to identify the individuals who will be implementing AI projects."
		13: "What we mostly need is programmers to create and integrate systems for process that can be simplified, and which later can be analysed accordingly to be made automatic."
		16: "Investing in people who have the right skills and retain resources at the government level ..."

(continued)

(continued)

Overarching theme 6: in demand		
Theme	Subtheme	Excerpts
	6.2.3 Agencies to the rescue	16: "… the teams are very onboard on embarking on EU-joint projects and attending international conferences where knowledge is diffused and exchange of experiences discussed, allowing for best practices to be discussed. It is very beneficial to learn from experts in the field."
		02: "…they assist the ministries in building prototypes to help shape the product and the requirements, but it's not the end result."
		10: "… pilot projects in the pipeline as proof of concept."
		13: "… systems adopted are pilot projects on their own and a first within government with the intention to explore technology and the possibility to improve how the public service operates."
	6.2.4 Mainstream education	14: "AI should not remain as a taboo topic and this could be done through our education system, ideally from the students at middle school level, and by educating and informing the public on AI concepts."
		13: "… education system, i would introduce one topic at a time such as process automation, through robotic courses that facilitates for fun learning, such as making a toy move."
		02: "The Maltese educational institutions are already offering very good programs, but also there are many other opportunities of financing to study abroad."

(continued)

(continued)

Overarching theme 6: in demand		
Theme	Subtheme	Excerpts
		17: "… scholarship is it's not just about taking an artificial intelligence program, but also a program in biomedical engineering or in health science, as long as it has some aspects of AI. This is important because technology, AI, is not just about creating it, it is multidisciplinary and we need to have all the different disciplines versed …"
		13: "… potentially encourage those students who are more inclined for these types of jobs and skills, engineers, scientists, programmers …"

Appendix K
Excerpts for Overarching Theme 7: AI as an Asset

Overarching theme 7: AI as an asset	
Theme	Excerpts
7.1 Enhancing processes	04: "Technology facilitation enhances workforce accountability and in cases, increases performance."
	09: "… introduction of AI systems to facilitate workflows for our staff."
	04: "Benefits are augmented as everything is reported at a click of a time, instant real-time reporting. Since we cannot increase the pay, we have to offer a level of flexibility and these systems provide us with a monitoring system that goes in hand with an increased level of accountability"
	07: "… facilitate work and get it done quicker …"
	10: "Improve overall health service delivery, productivity, and efficiency. Health professionals can shift from carrying out tedious and repetitive tasks to focus on other important tasks."
	16: "Proactiveness to automate with intelligence rather than reactive to situations …"
	15: "… biggest advantage of using these types of technologies is the provisioning of better, intelligent service …"
	04: "… we offer a service to ourselves and the biggest financial element that we have to look at is the human capital"
	01: "We will automate all that process and the impact will substantially result in efficacy gains of 30–70% and reduction of errors by 30–70%."
	14: "Minimise response timeframes from 40 to 4 s"
	15: "… there's a human that is supervising the teaching of the AI itself."

(continued)

M. Sciberras and A. Dingli, *Investigating AI Readiness in the Maltese Public Administration*, Lecture Notes in Networks and Systems 568,
https://doi.org/10.1007/978-3-031-19900-4

(continued)

Overarching theme 7: AI as an asset	
Theme	Excerpts
7.2 Effective use of resources	18: "We need to maximise the use of our public funds and do this effectively."
	11: "… the savings and operation efficiency …"
	13: "Performance is better than humans' in this regard."
	08: "… which for government will be an innovative way of tackling HR at the moment."
	01: "AI technology will need human intervention for the polishing, the monitoring, the regulation, the continuous review, because you can't just leave the autonomous system working on its own."
	10: "… support and possibly reduce further human errors."
7.3 Evidence based decisions	18: "… predictive AI is brilliant in drawing up patterns and pre-empting irregular and abnormal activity …"
	09: "Through streamlining, and the BI warehouse, the ministry can prepare for future projections …"
	06: "… benefits will outweigh the difficulties that this will entail essentially, an entire revolution of how our country and how the advanced societies would work."
	10: "… enhance the ability to identify certain genetic diseases … … increases the possibility of early detection of such diseases and helps in identifying the correct treatment procedures."
	10: "… patients better informed about the possible roots of their conditions and provide clinicians with better decision support."
	18: " This will facilitate the exploration of innovative concepts and data exploitation for evidence-based decision making"
	04: "We need to have all of the information systems plugged into a single system that will lead to informed decisions."
7.4 Investment	17: "… is attracting a number of different companies. What is the magic that we have to bring companies? I think one aspect is the access to resources, access to the individuals, access to it's very easy to get connections in Malta in regards to how to set up as there's a lot of support for that …"
	16: "Knowledge and human resources can attract foreign expertise and investment."
	18: "The size of the country, population density, the richness of our data in demographics, potentially places Malta in a very strong place for such potential investment."
7.5 Challenges	08: "… beneficial for everybody to see, you know, to eventually see these processes in place, but it really is time consuming …"
	08: "Engagement of resources, procurement and preserving knowledge are three huge challenges in the public service …"

Appendix L
Excerpts for Overarching Theme 8: Addressing Risks

Overarching theme 8: addressing risks	
Theme	Excerpts
8.1 Misuse of funds	17: "The risks, ultimately, are wasting public funds …"
	18: "This is of huge disappointment for technologists like me, who see a brilliant technological utility not exploited to bring about societal benefits …"
8.2 Mal-intent of AI solutions	10: "If misused, there can be a number of complications …"
	16: "… if something goes wrong there it will negatively reflect on Malta."
	18: "AI technology is badly labelled such as has happened to other innovative technologies due to deployments with malintent or malfunction."
8.3 Risk mitigation	04: "… and what the technical expectations are, we do mock-ups."
	10: "… pilot projects are being tried out, also so that we can address them in a regulatory framework."
	10: "… can mitigate certain things by having human resources to take final decision."
	16: "… because you're not investing a lot of effort, time and money into something which might fail. Working on proof of concepts at an applied level, I think it's very important …"
8.4 Identify accountability	08: "… reputation of the public service and the reputation of a service, the reputation of a minister, the reputation of the prime minister here at OPM, if we rely on AI to give us predictions about something, which then doesn't work out, what's the risk about that?"

(continued)

M. Sciberras and A. Dingli, *Investigating AI Readiness in the Maltese Public Administration*, Lecture Notes in Networks and Systems 568,
https://doi.org/10.1007/978-3-031-19900-4

(continued)

Overarching theme 8: addressing risks	
Theme	Excerpts
	10: "… a lot of factors that ultimately can impact on the decision taken by the machine or by the solution, which those consequences need to be addressed by the person responsible."
	18: "I am a firm believer that AI should aid humans and facilitate decision processes. I stop short of letting artificial intelligence take decisions instead of humans …"
	10: "… outcome that comes from an AI solution, it's not the definitive answer, meaning that so far professionals take the final decision."
	18: "Such systems can be programmed to act however I am of the strong belief that we have to further empower the human to take ultimate decisions."
8.5 Data quality	08: "… data quality one and legal challenges …"
	10: "… you don't have that quality, you can equally easily risk that certain decisions taken by this kind of technology to be either incorrect …"
	14: "… risks are dependent on the human factor as data input and the algorithms backing the system are all processed by humans."
	11: "… is biased algorithms; … Privacy and ethical issues."
8.6 Discontinuity of activity	17: "Continuity is very important. Once there is a plan that's accepted, keep pushing it."
	07: "… our current push is to bring it more to life, continue to process more data, which requests more human power …"
	16: "Another risk is the discontinuity due to lack of funds, disinterest or not enough vision to escalate the idea as individuals are focused on their own work or have different priorities."

Appendix M
Excerpts for Overarching Theme 9: Future Applications of AI

Overarching theme 9: future applications of AI	
Theme	Excerpts
9.1 Citizen centric	02: "… we are going to use the citizen chatbot of servizz.gov and build a data link to provide the citizen with a first level check of consumer complaint."
	15: "… using AI to make government more efficient and more citizen centric and more business centric is something which is key. The best projects will be those that are citizen centric, …"
	18: "driving forward initiatives that citizens can tangibly feel the positive changes AI will facilitate and I do that by working upon activities that are closest to the people."
	06: "… reach our end customers …"
	08: "… AI, the citizen twin concept where we have included that in the infrastructure architecture of our future public service, where we are hoping that what's the word, federated learning."
9.2 Image processing	01: "AI image processing and machine learning, hopefully we're going to be able to build a model that correctly identifies a valid …"
	02: "… we are looking at having a drone capturing anonymised information."
	07: "… cameras attached to vehicles, they can identify certain, vandalism …"
	07: "… basically most processes are image processing and machine learning with the data that is involved …"
	10: "AI for Medical Imaging"
	14: "Another AI solution, to enhance our geo-spatial data, will be connected through the global navigation satellite system."

(continued)

M. Sciberras and A. Dingli, *Investigating AI Readiness in the Maltese Public Administration*, Lecture Notes in Networks and Systems 568,
https://doi.org/10.1007/978-3-031-19900-4

(continued)

Overarching theme 9: future applications of AI	
Theme	Excerpts
	15: "It can be used for satellite imaging recognition and selection of illegal dumping …"
9.3 Analysis	11: "… AI to speed up search images, facilitate service specifications …"
	06: "… AI can be applied to risk analysis. In this regard I based the risk analysis software, I will include their machine learning process …"
	07: "… identifying them and then we're building a knowledge graph, …"
	08: "… some predictive analysis, from the data are predicting job opportunities of the future."
	11: "… predictive analytics and the use of big data generated by government to predict economic trends …"
	16: "… working on conversational bots …"
	15: "… do use of chat bots and bots to facilitate the communication in 24 by seven, without human intervention is very high on the agenda."
	06: "… we need to have these systems analysed for us, all this big data and give us at least a subset of data on which we can focus on, which we can concentrate."

Appendix N
Excerpts for Overarching Theme 10: Moving Forward

Overarching theme 10: moving forward		
Theme	Subtheme	Excerpts
10.1 Needs analysis		10: "… potentials of this technology are immense …"
		13: "Identify processes that can be standardised, procurement for example, and unify services."
		04: "That is where AI can be of great assistance, through analysing internal information to maximise the potential of the workforce."
		06: "… we need to ask what jobs will remain …"
		10: "An in-depth analysis is certainly required to address some of the issues …"
		11: "… group of people that represent the local, expertise even from the private sector, to see where the government should really invest and where savings can be generated when adopting AI technologies …"
		13: "Carry out a needs' analysis …"
		12: "… identify the services that can be quick wins across ministries, inform the public of these services and make effective use of the data."

(continued)

M. Sciberras and A. Dingli, *Investigating AI Readiness in the Maltese Public Administration*, Lecture Notes in Networks and Systems 568,
https://doi.org/10.1007/978-3-031-19900-4

(continued)

Overarching theme 10: moving forward		
Theme	Subtheme	Excerpts
10.2 Informative campaign		18: "For the public to understand and appreciate the work being done with AI, the awareness campaign for the citizens must be delivered in a way that the public can relate to it in a quotidian manner."
		05: "... education of the clients themselves, to make them accustomed with change ..."
		06: "I would start with educating the citizens, I will start by preparing them to what is going to happen ..."
		07: "... we need to put in more efforts to make it more visible."
		13: "Educate the public on AI in a relative manner, by relating to their everyday activities ..."
		10: "... educational campaigns about this technology are needed not only to our workforce as they will be the ones interacting on the forefront with this technology but also with customers ..."
		13: "It is a change process which we need to encourage through open and clear communication and for a positive and humane approach."
		13: "Set up a concise and clear communication strategy."
10.3 More to be done	10.3.1 Addressing the nation	13: "Education is important to instigate change, not just the within the current education system but also educating the public."
		09: "Education wise, I do not think we are reaching our objectives in the AI remit. I have discussed many a time that digital skills should be introduced earlier in our education system so that the future generations would compete better when it comes to top digital requirements."

(continued)

(continued)

Overarching theme 10: moving forward		
Theme	Subtheme	Excerpts
		16: "… education on all levels is important to ensure that we will adopt, we're not taking away jobs but train, to improve as a country, to improve from a knowledge perspective, and also try to be innovative and make our country more competitive in terms of attractiveness."
		16: "Education at all levels is important. When I say education, I think even a basic understanding of what AI can bring in terms of benefits. I think it's important from both a citizen point of view and from an employee point of view."
	10.3.2 Within the public administration	16: "… joint EU projects as they also transfer knowledge and will shed light on processes and systems that might be adaptive for local initiatives."
		17: "… further education as in bringing in the product owners to 1) really own the product and believe that it's their baby and see it forward and 2) educate them further on the importance of the technology, the importance of embracing this technology as a nation …"
		05: "… education, meaning I will have to make sure that the government has the right people who can read, interpret those data and that are also able to communicate the findings in a way that the final client would understand."
		10: "Encourage courses and scholarships to employees in the ICT Class."
		11: "Another aspect which needs to improve as a government as a whole is human resources in specialised fields, or training must be provided to identify the individuals who will be implementing AI projects."

(continued)

(continued)

Overarching theme 10: moving forward		
Theme	Subtheme	Excerpts
		07: "It must be an agency very specific, because as you know, there's already the MDIA which has AI within its remit but it is a national agency. It's not specifically addressing the internal public service AI requirements …"
		02: "Create an ecosystem, to champion AI and consider AI as a contributor to our economy."
		11: ".. best way is the top- bottom approach, which the central government should adopt an integrated holistic data strategy …"
		13: "Reform the ICT class across government."
		10: "… about the structure and the lack of it is very evident."
		06: "… I would set up a task force on AI on the adoption, of AI and then this task force, I would put people from different areas like IT, people have specialized in AI, psychologists, sociologists. There needs to be a multidisciplinary team to coordinate the future, which will be based on AI, because this will be a really radical shift, which its impact will be much more profound than it was the case with the internet."
		11: "… a central entity should delegate the ministries in this AI endeavour, as a one government holistic approach …"
		13: "… set the base guidelines on how to actually implement structures and give templates on how to process things."
		07: "… there should be a specific unit, it might be part of the CIO office, but I think there should be a specific unit in respect to specific skills required, dedicated to look into AI initiatives and focus on those AI initiatives, and obviously be able to collaborate even with other entities and ministries."

(continued)

(continued)

Overarching theme 10: moving forward		
Theme	Subtheme	Excerpts
		13: "What we mostly need is programmers to create and integrate systems for process that can be simplified, and which later can be analysed accordingly to be made automatic."
		16: "Investing in people who have the right skills and retain resources at the government level …"
	10.3.3 Increased collaboration	10: "… contact with MITA and Microsoft to identify suitable areas where AI can be introduced."
		03: "I believe that the University of Malta should collaborate with the government and present valuable research that can direct and assist in decision making. … More importantly, the UoM should present research and studies to the competent authorities for consideration."
		17: "… our primary role is as a regulator of technology, the secondary role is to support the government when it comes to digital innovation, emerging technology, policy input, …"
		16: "… joint EU projects as they also transfer knowledge and will shed light on processes and systems that might be adaptive for local initiatives."
		16: "I think competency in R&D and innovation are very important to create AI through business expertise; that is someone who understands AI, some who can work with data and some that has a better understanding of business. These three together, I think are very vital for a comprehensive approach."
		16: "Collaboration with the private sector, industrial partners and academic institutions is a key element when learning and implementing."

(continued)

(continued)

Overarching theme 10: moving forward		
Theme	Subtheme	Excerpts
		02: "… that attractive ecosystem, I think businesses will have the entrepreneurial spirit, they would automatically grasp the opportunity that they would automatically start creating economic activities around AI."
		03: "In its thousands of research studies conducted every year, the university should address real life problems when conducting research, it can be a collective effort. The government should trust the University of Malta with the research needed and not look elsewhere through external consultancy. The university has all the domains needed that can direct and give input while studies should be implemented and not end up on a shelf. The Maltese citizens should have an input on the future of the country and these studies can provide valuable insight; …"
		08: "Getting people more involved and the way the government is designing AI for services, so making people own it."
10.4 Tread cautiously		10: "More work is needed especially where it involves: data quality, ethics and regulatory frameworks …"
		06: "We are not weighing out, just looking at the benefits and we are not weighing out the actual impact of this massive introduction will have on the economy of the country, because at the end, it's the economy that is going to suffer the cost to have, I don't know, there is around 200 bus drivers and easily another 300 taxi drivers. That's 600 people, 600 of the out of job or out of business."
		17: "In relation to AI, I believe it's a bit too early to be loud about this …"

(continued)

(continued)

Overarching theme 10: moving forward		
Theme	Subtheme	Excerpts
		01: "AI would obviously replace humans. AI would replace taxi drivers, bus drivers, receptionists, the car sprayer, manufacturing and repetitive tasks. AI is replacing people today."

Bibliography

1. Aarseth W Rolstadas A, Andersen B (2013) Managing organisation challanges in global projects. Int J Manag Projects Bus 7(1)
2. Agrawal A, Gans J, Goldfarb A (2019) The economics of artificial intellignece: an agenda. The University of Chicago Press, Chicago
3. Ahmed A et al (2019) Knowledge-based systems survey. Int J Acad Eng Res 3:1–22
4. AI HLEG (2019) Ethics guidelines for trustworthy AI. https://digital-strategy.ec.europa.eu/en/library/ethics-guidelines-trustworthy-ai. Accessed 15 June 2021
5. Akdere M (2020) (Re)skilling employees for future work: how G20 countries can utilise artificial intelligence based learning technologies to scale up workplace training. https://www.g20-insights.org/policy_briefs/reskilling-employees-for-future-work-how-g20-countries-can-use-artificial-intelligence-based-learning-technologies-to-scale-up-workplace-training/. Accessed 10 May 2021
6. Akhtar N, Mian A (2018) Threat of adversarial attacks on deep learning in computer vision: a survey. IEEE Access 6:14410–14430
7. Alharbi YA, Alamoudi RH (2019) The role of technology in project management—a case study of Saudi company. IOSR J Eng 9(3):45–50
8. Allen G, Chan T (2017) Artificial intelligence and national security, 1st edn. Belfer center study, Cambridge, MA
9. Alpaydin E (2004) Introduction to machine learning. MIT Press, Massachusetts
10. Androutsopoulou A, Karacapilidis N, Loukis E, Charalibidis Y (2019) Transforming the communciation between citizens and governemnt through AI-guided chatbots. Gov Inf Q 36(2):358–367
11. APM (2020) What is project management?. https://www.apm.org.uk/resources/what-is-project-management/. Accessed 02 Mar 2021
12. Araya D (2019) Artificial intelligence and the end of government. https://www.forbes.com/sites/danielaraya/2019/01/04/artificial-intelligence-and-the-end-of-government/?sh=77ec71aa719b. Accessed 14 June 2021
13. Arntz M, Gregory T, Zierahn U (2016) The risk of automation for jobs in OECD countries: a comparative analysis. OECD, s.l.
14. Arvesen A, Hauan IB, Bolsoy BM, Hetwich EG (2015) Life cycle assessment of transport of electricity via different voltage levels: a case stusy for Nord-Trondelag county in Norway. Appl Energy 157:144–151
15. Ayodele TO (2010) Types of machine learning algorithms. New Adv Mach Learn 3:19–48

M. Sciberras and A. Dingli, *Investigating AI Readiness in the Maltese Public Administration*, Lecture Notes in Networks and Systems 568,
https://doi.org/10.1007/978-3-031-19900-4

16. Bacchi U (2021a) EU's lie-detecting virtual border guards face court scrunity. https://www.reuters.com/article/europe-tech-court-idUSL8N2KB2GT. Accessed 15 6 2021
17. Bacchi U (2021b) UPDATE 1-High-tech lie detector used at Europe borders face scrutiny. https://www.reuters.com/article/europe-tech-court-idUSL8N2KB346. Accessed 20 July 2021
18. Bai C, Dallasega P, Orzes G, Sarkis J (2020) Industry 4.0 technologies assessment: a sustainability perspective. Int J Prod Econ 229:107776
19. Bakker KD, Boonstra A, Wortmann H (2012) Risk managements' communicative effects influencing IT project success. Int J Project Manage 30:444–457
20. Bekey GA (2005) Autonomous robots: from biological inspiration to implementation and control, 2nd edn. Massachusetts Institute of Technology, Cmbridge, MA
21. Bekker S (2019) Fundamental rights in digital welfare states: the case of SyRI in the Netherlands. In Spijkers O, Werner WG, Wessel RA (eds) Netherlands yearbook of internation law. TMC Asser Press, The Hague, p vol 50
22. Belmas S, Akatov N, Kosyakin D, Belmas S (2021) Formation of the approach to assess the effectiveness of project initiation for company development. SHS Web of Conferences, 116(EDP Sciences), p 76.
23. Ben-Ari M, Mondala F (2018) Robots and their applications. Elements of robotics. Springer, Cham, pp 1–20
24. Bench-Capon TJ (1990) Knowledge representation: an approach to artificial intelligence. The A.P.I.C series, 32nd edn. Academic Press Ltd., London
25. Benessia A (2015) Sustainability and techno-science: what do we want to sustain and for whom? Int J Sustain Dev 18(4):329–348
26. Benjamins JW et al (2019) Enhancing cardiovascular artificial intelligence (AI) research in the Netherlands: CVON-AI consortium. Neth Heart J 9:414–425. https://doi.org/10.1007/s12471-019-1281-y
27. Bezzina F, Camilleri E, Marmara V (2021) Historical background of the Maltese public service administration and management (The turning point). Public service reforms in a small state island. Springer, Cham, pp 41–86
28. Bhattacharya S et al (2018) Open source platforms and frameworks for artificial inteeligence and machine learning. Institute of electrical and electronic engineers (IEEE), pp 1–4
29. Bhattacharya S et al (2018) Open source platforms and frameworks for artificial intellignce and machine learning. Institute of electrical and electronic engineers (IEEE), pp 1–4
30. Bini SA (2018) Artificial intelligence, machine learning, deep learning and cognitive computing: what do these terms mean and how will they impact health care? J Arthroplasty 33(8):2358–2361
31. Bogner A, Chanson M, Meeuw A (2016) A decentralised sharing app running a smart contract on the ethereum blockchain. In: Proceedings of the 6th international conference on the internet of things, pp 177–178
32. Bogoviz AV et al (2021) Frontier information technology and systems research in cooperative economics. In: Studues in systems, decision and control book 316. Springer, Switzerland AG
33. Bongardt D (2018) Germans have a critical attitude towards AI. https://www.hannovermesse.de/en/news/news-articles/germans-have-a-critical-attitude-towards-ai. Accessed 20 July 2021
34. Book TIP (2008) International form of contract, 1st edn. IChemE, Warwickshire
35. Bostrom N (2014) Superintelligence: paths, dangers, strategies. Oxford University Press, Oxford
36. Bova F, Goldfarb A, Melko R (2021) Quantum computing is coming. What can it do?. https://hbr.org/2021/07/quantum-computing-is-coming-what-can-it-do. Accessed 10 8 2021
37. Bovis C (2015) The law of EU public procurement. Oxford University Press, Oxford
38. Brachman RJ, Levesque HJ (2004) Knowledge representation and reasoning, 2nd edn. Elsevier, San Francisco, CA
39. Briguglio L (1988) The Maltese economy since 1960. https://www.um.edu.mt/library/oar/bitstream/123456789/24383/1/The%20Maltese%20economy%20since%201960.pdf. Accessed 19 July 2021

40. Broadbent E (2017) Interactions with robots: the truths we reveal about ourselves. Ann Rev Psychol 68:627–652. https://www.annualreviews.org/doi/pdf/10.1146/annurev-psych-010416-043958
41. Broadridge (2019) AI Adoption: Is your firm ready?. https://www.broadridge.com/_assets/pdf/broadridge-ai-readiness.pdf. Accessed 15 6 2021
42. Buckley JJ, Eslami E (2002) An introduction to fuzzy logic and fuzzy sets, 1st edn. Springer, Berlin
43. Bulkeley H, Newell P (2015) Governing climate change. Routledge, NY
44. Buxamann P, Hess T, Thatcher JB (2021) AI-based information systems. Bus Inf Syst 63:1–4
45. CAHAI-PDG (2021) Artificial intelligence in the public sector. Council of Europe, Strassbourg
46. Cambridge Dictionary (2021) Cambridge dictionary. https://dictionary.cambridge.org/dictionary/english/project. Accessed 02 Mar 2021
47. Capgemini Consulting (2018) Artificial intelligence benchmark. Capgemini Consulting, s.l.
48. Central Bank of Malta (2018) Development in the manufacturing sector. Central Bank Malta Q Rev 1:20–29
49. Cheng YW, Sun PC, Chen NS (2017) An investigation of the needs on educational robots. Timisoara, Romania, Institute of Electrical and Electronics Engineers, pp 536–538
50. Chen Y (2020) IoT, cloud computing, big data and AI in interdisciplinary domains. Simul Model Pract Theory 102(2020):1–5
51. Chidepatil A et al (2020) From trash to cash: how blockchain and multi-sensor driven artificial intelligence can transform circular economy of plastic waste. Admin Sci 10(2):23
52. Chircop G (1997) Evolution of a harbour infrastructure:from mercantile to naval control. Melita Historica 12(2):209–218
53. Chrissis MB, Konrad M, Shrum S (2011) CMMI: guidelines for process integration and product improvement. AddisonWesley, New York
54. Cimatti A, Pistore M, Traverso P (2008) Automated planning. In: van Harmelen F, Lifschitz B, Porter B (eds) The hand-book of knowledge representation. Elsevier B.V, Oxford, UK, pp 841–869
55. Coccia M (2020) Deep learning technology for improving cancer care in society: new directions in cancer imaging driven by artificial intelligence. Technol Soc 60:101–198
56. Collins (2021) Collins Dictionary. https://www.collinsdictionary.com/dictionary/english/issue. Accessed 30 May 2021
57. Compagnucci S et al (2020) The way to digital made-in-Europe. Promoting European values in the global digital arena. https://www.i-com.it/en/2020/12/08/the-way-to-digital-made-in-europe-promoting-european-values-in-the-global-digital-arena-2/. Accessed 4 May 2021
58. Cooke-Davies TJ, Arzymanow A (2003) The maturity of project management in different industries: an investigation into variations between project management models. Int J Project Manage 21(6):471–478
59. Council of Europe (2020) Guide on Article 8 of the European convention on human rights. https://www.echr.coe.int/documents/guide_art_8_eng.pdf. Accessed 08 June 2021
60. Creswell J, Plano CV (2007) Designing and conducting mixed methods research. Sage, Thousand Oaks, CA
61. Crown Commercial Services (2016) The public contracts regulations 2015 & The utilities contracts regulations 2016: guidance on awarding contracts. UK Government, London
62. Dantam NT, Kingston ZK, Chaudhuri S, Kavraki LE (2016) Incremental task and motion planning: a constraint-based approach. Robot Sci Syst 12: 52
63. Davenport T, Kalakota R (2019) The potential for artificial intelligence in healthcare. Future Healthc J 6(2):94
64. de Sousa WG et al (2019) How and where is artificial intelligence in the public sector going? A literature review and research agenda. Gov Inf Q 36(4):101392
65. Deloitte (2020) Covid-19 The upskilling imperative: building a future-ready workforce for the AI age. https://www2.deloitte.com/content/dam/Deloitte/ca/Documents/Analytics/ca-en-deloitte-analytics-upskilling-aoda.pdf: Canadian Financial Services

66. Deng L, Liu Y (2018) Chapter 1: a joint introduction to natural language processing and to deep learning. Deep learning in natural language processing. Springer, Singapore, pp 1–23
67. Desouza KC (2018) Delivering artificial intelligence in government: challanges and opportunities. IBM Centre for The Business of Government, Arizona, USA
68. Detelj K, Pihir I (2007) Influence of project management software tools usage on project success. University of Zagreb, Croatia
69. Deval V, Norta A (2019) Mobile smart-contract lifecycle governance with incentivized proof-of-stake for oligopoly-formation prevention. In: International symposium on cluster, cloud and grid computing, ppss 165–168
70. Diai Y et al (2020) Promoting students' well-being by developing their readiness for the artificial intelligence age. Sustainability 12(16):6597
71. Diaz VGP, Bonilla JPZ (2019) Exploring the effect of emerging technologies on scientific knowledge production and industrial advancement of society. In: Handbook of research on industrial advancement in scientific knowledge. IGI Global, s.l., pp 20–37
72. Dingli A, Fournier KS (2017) Financial time series forecasting—a machine learning approach. Proc Int Conf Mach Learn Appl 4(1/2/3):11–26
73. Dingli A, Haddod F, Kluver C (2021) Artificial intelligence in industry 4.0: a collection of innovative research case-studies that are reworking the way we look at Industry 4.0 thanks to artificial intelligence. Springer Nature Switzerland, Cham, Switzerland
74. Djeffal C (2020) Artificial intelligence and public governance: normative guidelines for artificial intelligence in government and public administration. Regulating artificial intelligence. Springer, Cham, pp 277–293
75. DW (2021) Sustainable mobility in cities. https://www.dw.com/en/sustainable-mobility-in-cities/av-50366795. Accessed 19 July 2021
76. DWIH-New York (2021) AI research taked root in Germany. https://www.dwih-newyork.org/en/current-focus-topics/artificial-intelligence/ai-research-takes-root-in-germany/. Accessed 20 July 2021
77. Edwards P, Bowen P (2013) Risk management in project organisations. Butterworth Heinemann, Oxford
78. Ekin A (2019) AI can help us fight climate change. But it has an energy problem, too. https://ec.europa.eu/research-and-innovation/en/horizon-magazine/ai-can-help-us-fight-climate-change-it-has-energy-problem-too. Accessed 14 June 2021
79. Engelmore RS, Feigenbaum E (1993) Expert systems and artificial intellingence. In: Engelmore RS (ed) Knowledge-based syetems in Japan. US Government, Maryland, pp 1–11
80. EU Commission (2019) 2030 climate & energy framework. https://ec.europa.eu/clima/policies/strategies/2030_en. Accessed 16 June 2021
81. EU Commission (2020) A European stratey for data. https://eur-lex.europa.eu/legal-content/EN/TXT/?qid=1593073685620&uri=CELEX%3A52020DC0066. Accessed 06 May 2021
82. EU Commission (2020) Regulation of the European Parliment and of the Counsil on European Data Governance. COM(2020) 767 final, Brussels
83. EU Commission (2020) White paper on artificial intelligence—A European approach to excellence and trust. https://eur-lex.europa.eu/legal-content/EN/TXT/?uri=CELEX%3A52020DC0065&WT_mc_id=Twitter. Accessed 06 May 2021
84. EU Parliment (2020) Artificial intelligence and how it is used?. https://www.europarl.europa.eu/news/en/headlines/society/20200827STO85804/what-is-artificial-intelligence-and-how-is-it-used. Accessed 12 June 2021
85. European Parliment (2021) Artificial intellignece: threats and opportunities. https://www.europarl.europa.eu/news/en/headlines/society/20200918STO87404/artificial-intelligence-threats-and-opportunities. Accessed 02 May 2021
86. Evas T, Heflich A (2021) Artificial intelligence in road transport. European Parlimentary Research Service, Brussles
87. FCT (2021) R&D Projects. https://www.fct.pt/apoios/projectos/concursos/datascience/index.phtml.en. Accessed 20 July 2021

88. Feikert-Ahalt C (2018) Malta: government passes three law to encourage blockchain technology. https://www.loc.gov/item/global-legal-monitor/2018-08-31/malta-government-passes-three-laws-to-encourage-blockchain-technology/. Accessed 16 June 2021
89. Feldstein S (2019) The global expansion of AI surveillance, vol 17. Carnegie Endowment for International Peace, Washington DC
90. Felix CV (2020) The role of the teacher and AI in education. In: Sengupta E, Blessinger P, Makhanya MS (eds) International perspectives on the role of technology in humanizing higher education. Bingley, Emerald Publishing, pp 33–48
91. Ferentinos KP (2018) Deep learning models for plant disease detection and diagnosis. Comput Electron Agric 145:311–318
92. Ferrus RM, Somonte MD (2016) Design in robotics based in the voice of the customer of household robots. Robot Auton Syst 79:99–107
93. Fiksel J (2015) From risk to resilience. In: Washington DC (ed) Resilience by design. Island Press, pp 19–34
94. Finchman RG, Dos Santos BL, Zheng Z (2014) Digital innovation as a fundamental and powerful concept in the information systems curriculum. MIS Q 38(2):329–353
95. Fink D (2014) Project risk governance: managing uncertainty and creating organisational value. Gower Publishing Ltd., Surrey
96. FinlandWorks (2021) Artificial intelligence from Finland. https://toolbox.finland.fi/wp-content/uploads/sites/2/2021/01/bf_ai_from_finland_web.pdf. Accessed 19 July 2021
97. Floridi L (2020) Artificial intelligence as a public service: learning from Amsterdam and Helsinki. Philos Technol 33(4):541–546
98. Francis D, Horine G (2004) PMP: exam cram 2, 1st edn. Pearson IT, s.l.
99. Freese C, Schalk R (1996) Implications of differences in psychological contracts for human resource management. Eur J Work Organ Psychol 5(4):501–509
100. Frendo H (1988) The British Colonial Experience 1800–1964: impact on the Maltese society. In: Mallia-Milanes V (ed) Maltese Colonial identity: Latin Mediterranean or British Empire?. Mireva Productions, Msida, pp 185–214
101. Galea C (2006) A better connected hi-tech malta. https://timesofmalta.com/articles/view/a-better-connected-hi-tech-malta.65462. Accessed 16 June 2021
102. Galizia PC (2017) The economy of modern Malta: from the nineteenth to the twenty-first century. In: Deng K (ed) Palgrave studies in economic history. Spinger Nature, NY, pp 1–391
103. Garrett CR, Lonzano-Perez T, Kaelbling LP (2018) FFRob: leveraging symbolic planning for efficient task and motion planning. Int J Robot Res 37(1):104–136
104. Gartner (2018) Gartner survey of more than 3,000 CIOs reveals that enterprises are entering a third era of IT. https://www.gartner.com/en/newsroom/press-releases/2018-10-16-gartner-survey-of-more-than-3000-cios-reveals-that-enterprises-are-entering-the-third-era-of-it. Accessed 12 May 2021
105. Ghallab M, Nau D, Traverso P (2004) Automated planning: theory and practice. Elsevier, San Francisco, CA
106. Ghallab M, Nau D, Traverso P (2016) Automated planning and acting, 1st edn. Cambridge University Press, NY
107. Ghobakhloo M (2020) Industry 4.0: digitisation and opportunities for sustainability. J Clean Prod 252:119869
108. Giampaoli D, Ciambotti M, Bontis N (2017) Knowledge managements, problem solving and performance in top Italian firms. J Knowl Manag 21(2):355–375
109. Gido J, Clements J (2014) Successful project management. Cengage Learning, Boston
110. Goertzel B, Pennachin C (2007) Artificial general intelligence. Springer, Berlin
111. Gonzi RD (2019) Change and continuity management in the public sector: the DALI model for effective decision making. Emerald Group Publishing, Bingley
112. Government of Ireland (2021) AI-here for good: a national artificial intelligence strategy for Ireland. https://www.gov.ie/en/publication/91f74-national-ai-strategy/. Accessed 3 8 2021
113. Grace K et al (2018) When will AI exceed human performance? Evidence from AI experts. J Artif Intell Res 62:729–754

114. Grech AG (2015) The evolution of the Maltese economy since independence. Central Bank of Malta, Valletta
115. Grech AG (2015) Understanding the macroeconomic impact of migration in Malta. Central Bank of Malta, Valletta
116. Grech AG (2017) Did Malta's accession to the EU raise its potential growth? A focus on the foreign workforce. J Econ Integr 32(4):873–890
117. Griep Y, Cooper C (2019) Handbook of research on the osychological contract at work, 1st edn. Edward Wlgar, Cheltenham, UK
118. Haenssle HA et al (2018) Man against machine: diagnostic performance of a deep learning convolutional neural network for dermoscopic melanoma recognision in comparison to 58 detmatologists. Ann Oncol 29:1836–1842
119. Harwood M (2020) From Euroscepticism to Euro-enthusiasm: how Malta weathered the eurozone crisis. https://www.tandfonline.com/doi/full/10.1080/13608746.2020.1746542. Accessed 19 July 2021
120. Haumann S (2018) Towards a historical understanding of critical raw materials. Gaia 27(4):373–378
121. Hawkswort J, Berriman R, Cameron E (2018) Will robots really steal our jobs? An international analysis of potentail long term impact of automation. PWC, UK
122. Haynes K (2012) Reflexivity in qualitative research. In: Symon G, Cassell C (eds) Qualitative organizational research: core methods and current challenges. SAGE Publications Inc., London, pp 72–89
123. Helmstaedter M et al (2013) Connectomic reconstruction of the inner plexiform layer in the mouse retina. Nature 500(7461):168–174
124. Hengstler M, Enkel E, Duelli S (2016) Applied artificial intelligence and trust-the case of autonomous vehicles and medical assistance devices. Technol Forecast Soc Change 105:105–120
125. High R (2012) The era of cognitive systems: an inside look at IBM Watson and how it works. IBM Watson, s.l.
126. Hough B, David H (2009) The British right to rule Malta, 1800–1813. Melita Historica 14(4):387–408
127. Hrastinski S et al (2019) Critical imaginaries and reflections on artificial intelligence and robots in postdigital K-12 education. Postdigit Sci Educ 1(2):427–445
128. Huang RH et al (2020) Handbook on facilitating flexible learning during educational disruption: the Chinese experience in maintaining undisrupted learning in COVID-19 outbreak. Smart Learning Institute of Beijing Normal University, Beijing, vol 1.2, pp 1–54
129. Huan TD et al (2018) A universal startegy for the creation of machine learning-based atomistic force fields. NPJ Comput Mater 3(1):1–8
130. Hunt VD (1996) Process mapping: how to re-engineer your business processes. Wiley & Sons, New Jersey
131. IBQMI (2015) International business & quality management institute. https://news.ibqmi.org/five-challenges-faced-by-project-managers. Accessed 03 July 2021
132. INCoDe, (2019) AI Portugal 2030: Portuguese national initiaitive on DIGI. Fundação para a Ciência e Tecnologia, Lisbon
133. Indrasen P (2017) How artificial intelligence is impacting the real life everyday. Int J Adv Res Dev 2(10):96–100
134. ISACA (2009) The risk IT framework based on CobiT. ISACA, IL, USA
135. Is-Servizz Pubbliku (2021) Is-Servizz Pubbliku. https://www.facebook.com/servizzpubbliku. Accessed 25 June 2021
136. Ivankova NV, Creswell JW, Stick SL (2006) Using mixed-methods sequential explanatory design: from theory to practice. Field Methods 18(1):3–20
137. Ivanov S, Webster C (2018) Adoption of robots, artificial intelligence and service automation by travel, tourism and hospitality companies—A cost benefit analysis. In: Marinov V, Vodenska M, Assenova M, Dogramadjieva E (eds) Traditions and innovations in contemporary tourism. Newcastle upon Tyne, Cambridge, pp 190–203

138. Jensen AP, Voordt TV (2017) Facilities management and corporate real estate management as value drivers, 2nd edn. Routledge, New York
139. Jha KN (2011) Construction project management: theory and practice. Pearson, Delhi
140. Jones D, Savage D, Westgate R (2003) Partnering and collaborative working: law and industry practice. Informa Law from Routledge, Routledge
141. Jonsen K, Jehn KA (2009) Using triangulation to validate themes in qualitative studies. Qual Res Organ Manage Int J 4(2):123–150
142. JRC (2020) AI watch: artificial intelligence in the public services. Luxembourg publications office of the EU, Luxembourg
143. Junior RR, Carvalho MM (2013) Understanding the impact of project risk management on project performance: an empirical study. J Technol Manage Innov 8(ALTEC):64–78
144. Kaluarachchi T, Reis A, Nanayakkara S (2021) A review of recent deep learning approaches in human centered machine learning. Sensors 21(7):2514
145. Kaluarachchi T, Reis A, Nanayakkara S (2021) A review of recent deep learning approaches in human-centered machine learning. Sensors 21(7):2514
146. Kanapinskas V, Plytnikas Z, Tvaranaviciene A (2014) Sustainable public procurement: realization of the social aspect in Republic of Lithuania. Bus Theory Pract 15(4):302–315
147. Kaneko Y, Yada K (2016) A deep learning approach for the prediction of retail stores sales. In: 2016 IEEE: 16th international conference on data mining workshops (ICDMW), pp 531–537
148. Kashta R (2014) Corruption and innovation in the Albanian public procurement system. Int Sci J 2:212–225
149. Khan O, Burnes B (2007) Risk and supply chain management: creating a research agenda. Int J Logist Manage 18(2):197–216
150. Kim EH, Oh SK, Pedrycz W (2017) Reinforced rule-based fuzzy models: design and analysis. Knowl-Based Syst 119:44–58
151. Kononenko I, Kharazii A (2014) The methods of selection of the project management methodology. Int J Comput 13(1):1–2
152. Kopponen A, Ruostetsaari N (2021) Implementation of the national AuroraAI programme. https://vm.fi/en/auroraai-en. Accessed 19 July 2021
153. Kordjamshidi P, Moens MF (2015) Global machine learning for spatial ontology population. J Web Semant 30:3–21
154. KPMG (2013) Self-driving cars: are we ready? KPMG, USA
155. Krulik S, Rudnick JA (1987) Problem solving: a handbook for teachers, 2nd edn. Allyn and Bacon, Boston
156. Krystal M (2020) The upskilling imperative: building a future-ready workforce for the AI age. Deloitte, Canada
157. Kugler L (2018) AI judges and juries. Commun ACM 61(12):19–21
158. Kwaka R et al (2011) Metropolitan areas in Europe. https://inta-aivn.org/images/cc/Metropolisation/background%20documents/Metropolitan_Europe_BBSR_Study.pdf. Accessed 16 June 2021
159. Lakshmi KJ, Kumar TK, Warrier S (2021) Automated Face recognition by smart security secutiry systems using AI & ML algorithms. In: 5th international conference on trends in electronics and informatics (ICOEI), volume IEEE, pp 1363–1368
160. Lamichhane CD (2018) Understanding the education philosophy and its implications. NCC (Nepal Commerce Campus) J 3(1):24–29
161. Lamichhane CD (2018) Understanding the education philosophy and its implincations. NCC (Nepal Commerce Campus) J 3(1):24–29
162. Li J et al (2019) Combined sensing, cognition, learning and control for developing future neuro-robotics systems: a survey. Inst Electr Electron Eng Trans Cogn Dev Syst 11(2):148–161
163. Lim TJ (2020) How Finland is using AI for predictive public services. https://govinsider.asia/digital-economy/how-finland-is-using-ai-for-predictive-public-services/. Accessed 01 July 2021
164. Lips-Wiersma M, Wright S, Dik B (2016) Meaningful work: differences among blue-, pink-, and white-collar occupations. Career Dev Int 21(5):534–552

165. Liu HC, You JX, Tian G (2017) Fuzzy Petri nets for knowledge representation and reasoning: a literatre review. Eng Appl Artif Intell 60:45–56
166. Loce RP, Bernal EA, Wu W, Bala R (2013) Computer vision in roadway transportation systems: a survey. J Electron Imag 22(4):041121
167. Luckin R, Holmes W, Griffiths M, Forcier LB (2016) Intelligence unleashed: an argument for AI in education. http://oro.open.ac.uk/50104/1/Luckin%20et%20al.%20-%202016%20-%20Intelligence%20Unleashed.%20An%20argument%20for%20AI%20in%20Educ.pdf. Accessed 15 June 2021
168. Machado CG et al (2021) Digital organisational readiness: experiences from manufacturing companies. J Manuf Technol Manag 32(9):167–182
169. Mackieson P, Shlonsky A, Connolly M (2019) Increasing rigor and reducing bias in qualitative research: A document analysis of parliamentary debates using applied thematic analysis. Qual Soc Work 18(6):965–980
170. Magazine M (1914) British rule in Malta. https://www-proquest-com.ejournals.um.edu.mt/magazines/british-rule-malta/docview/1437742097/se-2?accountid=27934. Accessed 16 June 2021
171. Mahadevan SI (2018) Imagination machines: a new challenge for artificial intelligence. AAAI 32(1):7988–7993
172. Makridakis S (2017) The forthcoming artificial intelligence (AI) revolution: its impact on society and firms. Futures 90:46–60
173. Manyika J et al (2011) Big Data: the next frontioer for innovation, competition and productivity. McKinsey Global Institute, Washington, District of Columbia
174. Manyia J et al (2017) Jobs lost, jobs gained: what the future of work will mean for jobs, skills, and wages. McKinsey Global Institute, s.l.
175. Markovitz G (2020) At least half of the people who have a job fear they'll lose it in the next 12 months. https://www.weforum.org/agenda/2020/10/more-than-half-of-working-adults-fear-for-their-jobs/. Accessed 15 June 2021
176. Marr B (2018) What is Industry 4.0? Here is a super wasy explanation for anyone. https://www.forbes.com/sites/bernardmarr/2018/09/02/what-is-industry-4-0-heres-a-super-easy-explanation-for-anyone/?sh=2e3dd48f9788. Accessed 02 May 2021
177. Mccarthy JR, Holland J, Gillies V (2003) Multiple perspectives on the 'family lives' of young people: methodological and theoretical issues in case study research. Int J Soc Res Method 6(1):1–23
178. McGovern A et al (2017) Using artificial intelligence to improve real-time decision making for high-impact weather. Bull Am Meteorol Soc 98(10):2073–2090
179. McKendrick E (1997) Capacity in contract law, 1st edn. Palgrave, London
180. McKendrick K (2019) Artificial intelligence prediction and counterterrorism. Chatham House, London
181. McKenna L, Gray R (2018) The importance of ethics in research publications. Collegian 25(2):147–148
182. McKinsey (2017) A future that works: automation, employment and productivity. McKinsey Global Institute, San Francisco
183. McKinsey (2018) AI, automation, and the future of work: ten things to solve for. McKinsey Global Institute, s.l.
184. McLaughlin SA (2017) Dynamic capabilities: taking an emerging technology perspective. Int J Manuf Technol 31:62–81
185. McManamny T, Sheen J, Boyd L, Jennings PA (2015) Mixed methods and its application in prehospital research: a systematic review. J Mixed Methods Res 9(3):214–231
186. McNeal GS, Goodwin W, Jones S (2017) Warrantless operations of public use drones: Considerations for government agencies. Fordham Urb L J 44:703–723
187. MDIA (2020) Malta digital innovation authority. https://mdia.gov.mt/maltas-national-ai-strategy-ranked-10th-best-amongst-54-countries/. Accessed 01 Mar 2021
188. MDIA (2020) Malta's national AI strategy ranked 10th best amongst 54 countries. https://mdia.gov.mt/maltas-national-ai-strategy-ranked-10th-best-amongst-54-countries/. Accessed 16 June 2021

189. Mekki K, Baijic E, Chaxel F, Meyer F (2019) A comparative study of LPWAN technologies for large scale IoT deployment. ICT Express 1(5):1–7
190. Mesko B (2017) The role of artificial intelligence in precision medicine. Exp Rev Precis Med Drug Dev 2(5):239–241
191. Meyer AD, Loch CH, Pich MT (2002) managing project uncertainty: from variation to chaos. MIT Sloan Manage Rev 43:60+
192. Mhlanga D (2020) Industry 4.0 in finance: the impact of artificial intelligence on digital financial inclusion. Int J Financ Stud 8(3):45
193. Miailhe N, Hodes C (2017) The third age of artificial intelligence. Fields Actions Sci Rep 17(Special Issue):6–11
194. Microsoft (2020) Artificial intelligence in the public sector—European outlook for 2020 and beyond. https://info.microsoft.com/WE-DTGOV-CNTNT-FY21-09Sep-22-ArtificialIntelligenceinthePublicSector-SRGCM3835_01Registration-FormінBody.html. Accessed 21 May 2021
195. Miller D (2019) The medical AI insurgency: what physicians must know about data to procatice with intelligent machines. NPJ Digit Med 2(62):1–5
196. Misuraca H, Van Noordt C (2020) AI watch: artificial intelligence in public services. Publication office of the European Union, Luxembourg
197. Montes-Guerra MI, Gimena FN, Perez-Ezcurdia MA, Diez-Silva HM (2014) The influence of monitoring and control on project management success. Int J Constr Proj Manage 6(2):163–184
198. Muggleton S (2014) Alan Turing and the development of artificial intelligence. Al Commun 27(1):3–10
199. Muro M, Maxim R, Whiton J (2019) Automation and artificial intelligence: how machines are affecting people and places. Metropolitan Policy Program by Brookings, Washington
200. Musen MA (2015) The protege project: a look back and a look forward. AI Matters 1(4):4–12
201. Najdawi A (2020) Assessing AI readiness across organisations: the case of UAE. In: 2020 11th international confernece on computing, communication and networking technologies. IEEE, vol 49239, pp 1–5
202. Narejo GB (2020) AI in healthcare. In: Agarwal S, Makkar S, Tran D (eds) Privacy vulnerablites and data security challenges in the IoT. CRC Press, s.l., p 16
203. Nguyen HT, Walker C, Walker EA (2018) A first course in fuzzy logic, 4th edn. Chapman and Hall/CRC, NY
204. OECD.AI (2021a) iSIMPLEX. https://www.oecd.ai/dashboards/policy-initiatives/http:%2F%2Faipo.oecd.org%2F2021-data-policyInitiatives-25703. Accessed 20 July 2021
205. OECD.AI (2021b) National AI policies and strategies. https://www.oecd.ai/dashboards. Accessed 20 July 2021
206. OECD (2021) Chapter 13: case study 8: Estonia e-government and the creation of a comprehensive data infrastructure for public servicesand agriculture policies implementation. https://www.oecd-ilibrary.org//sites/510a82b5-en/index.html?itemId=/content/component/510a82b5-en#section-d1e19174. Accessed 20 July 2021
207. Oh M, Choi S (2020) The competence of project team members and success factors with open innovation. J Open Innov 6(51):1–16
208. Oleshchuk V, Fensli R (2011) Remote patient monitoring within a future 5G infrastructure. Wirel Pers Commun 57:431–439
209. OPSI (2021) UNA—The first virtual assistant of public administration in Latvia. https://oecd-opsi.org/innovations/una-the-first-virtual-assistant-of-public-administration-in-latvia/. Accessed 20 July 2021
210. Osborne SP, Radnor Z, Strokosch K (2016) Co-production and the co-creation of the value in public services: a suitable case for treatment? Public Manag Rev 18(5):639–653
211. Oxborough C, Ringger T, Versage MS (2019) Three governance considerations to unlock the potential of AI. https://www.digitalpulse.pwc.com.au/ai-governance-considerations/. Accessed 30 June 2021

212. Oyedeji A et al (2019) Design and implementation of a medical diagnostic expert system. J Eng Sci 10:103–109
213. Patricio DI, Reider R (2018) Computer vision and artificial intelligence in precision agriculture for grain crops: a systematic review. Comput Electron Agric 153:69–81
214. Patterson J, Gibson A (2017) deep learning: a practitioner's approach, 1st edn. O'Reilly Media, Sebastopol, CA
215. Paucar C et al. (2018) Use of drones for surveillance and reconnaissance of military areas. In: International conference of research applied defense and security, pp 119–132
216. Pencarelli T (2019) The digital revolution in the travel and tourism industry. Inf Technol Tour 22(3):455–476
217. Peplow R (2016) The progression form eGovernment to mGovernment. Univeristy of Malta, Msida
218. Pereira AA, Espada JP, Crespo RG, Aguilar SR (2019) Platform for controlling and getting data from network connected drowns in indoor environments. Futur Gener Comput Syst 92:656–662
219. Perez-Fernandez D et al (2019) Corpus viewer: NLP and ML-based platform for public policy making and implementation. Procesamiento del Lenguaje Natiral 63:193–196
220. Perminova O, Gustaffson M, Wikstorm K (2008) Defining uncertainty in projects-a new perspective. Int J Project Manage 26:73–79
221. Perzylo A et al (2019) SMErobotics: smart robots for flexible manufacturing. IEEE Robot Autom Mag 26(1):78–90
222. PMBoK (2017) A guide to the project management body of knowledge, 6th edn. Project Management Institute, Pennsylvania
223. PMI (2021) Project management institute. https://www.pmi.org/about/learn-about-pmi/what-is-project-management. Accessed 03 July 2021
224. Polygant (2021) Artificial Intelligence will become part of the public administration system in Latvia. https://polygant.net/blog/artificial-intelligence-will-become-part-of-the-public-administration-system-in-latvia/. Accessed 20 July 2021
225. Portugal INCoDe 2030 (2019) AI Portugal 2030. https://www.incode2030.gov.pt/en/documents-studies. Accessed 18 6 2021
226. Putterman L (2008) Agriculture, diffusion and development: ripple effects of the neolithic revolution. Economica 75:729–748
227. PwC (2019) Malta's new regulatory framework built for blockchain technology. https://www.pwc.com/mt/en/publications/technology/pwc-malta-blockchain-alert.html. Accessed 16 June 2021
228. PwC (2021) A leadership agenda to take on tomorrow. https://www.pwc.com/gx/en/ceo-agenda/ceosurvey/2021.html?sf55546653=1%C2%A0%E2%80%A6. Accessed 16 June 2021
229. Rael R (2017) Strategy and risk management: an integrated practical approach. American Institute of Certified Public Accountants Inc., New York
230. Raji ID, Buolamwini J (2019) Actionable auditing: investigating the impact of publicly naming biased performance results of commercial products. Honolulu, USA, s.n.
231. Ranjan A, Behera V, Reza M (2021) OCR using computer vision and machine learning. In: Das SK, Das SP, Dey N, Hassanien AE (eds) Machine learning algortihms for industrial applications. Springer, s.l., pp 83–105
232. Rawassizadeh R et al (2019) Manifestation of virtual assistants and robots into daily life: vision and challenges. CCF Trans Pervasive Comput Interact 1:163–174
233. Ray C, Mondada F, Siegwat R (2008) Whst to people expect from robots?. In: International conference on intelligent robots and systems, pp 3816–3821
234. Rehman A, Saba T (2014) Evaluation of artificial intelligence techniques to secure information in enterprises. Artif Intell Rev 42(4):1029–1044
235. Reis J, Santo PE, Melao N (2019) Impacts of artificial intelligence on public administration: a systematic literature review. In: 14th Iberian conference on information systems and technologies (CISTI), Coimbra, Portugal

236. Renn O (2014) Emerging risks: methodology, classification and policy implications. J Risk Anal Crisis Response 4(3):114–132
237. Richard OC, Mc-Millan-Capehart A, Bhuian SN, Taylor EC (2009) Antecedents and consequences of psychological contracts: does organisatioanl culture really matter? J Bus Res 62(8):818–825
238. Rittingjouse JW, Ransome JF (2017) Cloud computing: implementation, management and security, 1st edn. CRC Press, Boca Raton
239. Ross PT, Zaidi NL (2019) Limited by our limitations. Perspect Med Educ 8(4):261–264
240. Rousseau DM, Schalk R (2000) Psychological contracts in employment: cross-national perspectives, 1st edn. Sage Publications, London
241. Russell S, Dewey D, Tegmark M (2015) Research priorities for robust and beneficial artificial intelligence. In: Association for the advancement of artificial intelligence, vol Winter, pp 105–115
242. Sahu P, Dua M, Kumar A (2018) Challenges and issues in adopting speech recognition. In: Agrawal S, Devi A, Watson R, Bansal P (eds) Speech and language processing for human-machine communications. Advances in intelligent systems and computing, vol 664. Springer, Singapore, pp 209–215
243. Samoilo S et al (2020) AI watch: defining artificial intelligence. Publications Office of the European Union, Luxemburg
244. Sanchez-Pi N, Marti L (2021) Towards a green AI: evolutionary solutions for an ecologically viable artificial intelligence. In: Proceedings of the genetic and evolutionary computatuin conference companion, pp 1135–1140
245. Santra D, Basu SK, Mandal JK, Goswami S (2020) Rough set based lattice structure for knowledge representation in medical expert systems: low back pain management case study. Exp Syst Appl 145(113084)
246. Saravi S et al (2019) Use of artificial intelligence to improve resilience and preparedness against adverse flood events. Water 11(5):973
247. Schmidt E et al. (2021) National security commission on artificial intelligence. USA National Security Commission, s.l.
248. Scott I, Gong T (2021) Coordinating government dilos: challanges and opportunites. GPPG 1:20–38. https://doi.org/10.1007/s43508-021-00004-z
249. Segal E (2021) The impact of AI on cybersecurity. https://www.computer.org/publications/tech-news/trends/the-impact-of-ai-on-cybersecurity. Accessed 15 June 2021
250. Segal S (2019) Corporate value of enterprise risk management. Wiley & Sons Inc., New Jersey
251. Serholt S (2017) Child-robot interaction in education. Gothenburg University, Gothenburg
252. Servoz M (2021) The future of work? Work of the future! On how artificial intelligence, robotics and automation are transforming jobs and economy in Europe. EU Commission, s.l.
253. Seyour T, Hussein S (2014) The history of project management. Int J Manag Inf Syst 18(4):233–240
254. Shah H, Warwick K, Vallverdu J, Wu D (2016) Can machines talk? Comparison of Eliza with modern dialogue systems. Comput Hum Behav 58:278–295
255. Sharkey AJ (2016) Should we welcome robot teachers? Ethics Inf Technol 18(4):283–297
256. Sigala M, Rahimi R, Thelwall M (2019) Big data and innovation in tourism, travel and hospitality: managerial apporahces, techniques and applications. Springer Nature SIngapore Pte Ltd., SIngapore
257. Smith BC (2019) The promise of artificial intelligence: reckoning and judgement, 1st edn. MiT Press, Cambridge, MA
258. Smith JC et al (2012) Projected demand and potential impacts to the mational airspace system of autonomous, electric, on-demand small aircraft. Aerospace Research Centre, Indianapolis, Indiana
259. Smith M, Miller S (2021) The ethical application of biometric facial recognition technology. Ai & Society, pp 1–9
260. Spruds A, Potjomkina D, Scerbinskis V (2017) Activities and personalities. In: The centenary of Latvia's foreign affairs. Latvian Institute of International Affairs, Riga, p 288

261. Starns VA (2019) Exploring the strategies project managers need to establish a project charter for initiating a project. Colorado Technical University, Colorado
262. Stiftung B (2021) Automating society report 2020. https://automatingsociety.algorithmwatch.org/report2020/portugal/. Accessed 20 July 2021
263. Sudha C, Raj TN (2017) Analysis of suspicious pattern discovery using AI-neural network in credit card fraud detection. IJCRR 9(10):80–83
264. Suresh GA, Kumar AS, Lekashri S, Manikandan R (2021) Efficient crop yield recommendation system using machine learning for digital farmin. Int J Mod Agric 10(1):906–914
265. Susanti A, Soemitro RA, Suprayitno H, Ratnasari V (2019) Searching the appropriate minimum sample size calculation method for computer train passenger travel behavior survey. J Infrastruct Facil Asset Manage 1(1):47–61
266. Tabassi AA, Bryde DJ, Kamal EM, Dowson J (2018) Challenges for project management in the 21st century. Malaysia, 7th ICBEDC-21–23 Nov
267. Taeihagh A, Lim HS (2019) Governing autonomous vehicles: emerging responses for safety, liability, privacy, cybersecurity and industry risks. Transp Rev 39(1):103–128
268. Tayal DK et al (2015) Crime detection and criminal identification in India using data mining techniques. AI & Soc 30(1):117–127
269. Terzieva M (2014) Project knowledge management; how organisations learn from experience. Elsevier, Milan, Italy
270. Thamhain H (2013) Managing risks in complex projects. Proj Manage J 44(2):20–35
271. Theofanidis D, Fountouki A (2018) Limitations and delimitations in the research process. Periop Nurs 7(3):155–163
272. Thorpe CP, Bailey JC (1999) Commercial contracts: a practical guide to deals, contracts, agreements & promises, 2nd edn. Kogan Page, London
273. Tian X, Slocum JW (2016) Managing corporate social responsibility in China. Organ Dyn 45(1):39–46
274. Tolun MR, Oztoprak K, Sahin S (2016) Expert systems. In: Kirk-Othmer (ed) Encyclopedia of chemical technology. Wiley, s.l.
275. Tyagi AK, Chahal P (2020) Challenges and applications for implementing machine learning. In: Kashyap R, Kumar AV (eds) Computer vision. Advances in computer and electrical engineering. IGI Global, s.l., pp 188–220
276. UK Parliment (2020) AI and Healthcare. https://researchbriefings.files.parliament.uk/documents/POST-PN-0637/POST-PN-0637.pdf. Accessed 14 June 2021
277. UN (2015) Paris agreement. UN, Paris
278. UNEP (2021) Closing the gap. https://www.unep.org/resources/factsheet/closing-gap. Accessed 15 June 2021
279. Urban G et al (2018) Deep learning localises and identifies polyps in real time with 96% accuracy in screening colonoscopy. Gastroenterology 155:1069–1078
280. van Noordt C, Misuraca G (2019) New wine in old bottles: Chatbots in government. Electron Participation 11686:49–59
281. Vance CM, Paik Y (2011) Managing a global workforce: challanges & opportunities in international human resource management, 2nd edn. Routledge, London
282. Vassallo M (2006) Malta: "Feeding in" and "Feeding out", and integrating immigrants and ethnic minorities: a study of national policies. European Commission, Brussels
283. Vincent-Lancrin S, Urgel J, Kar S, Jacotin G (2019) Measuring innovation in education 2019: what has changed in the classroom? OECD, s.l.
284. Voulodimos A, Doulamis N, Doulamis A, Protopapadakis E (2018) Deep learning for computer vision: a brief review. Comput Intell Neurosci 2018(Article ID 7068349):1–13
285. Vyborny CJ, Giger ML (1994) Computer vision and artificial intelligence in mammography. Am J Roentgenol 162(3):699–708
286. Webb M (2020) The impact of artificial intelligence on the labor market. Stanford University, Stanford
287. WEF (2020) The future of jobs report 2020. WEF, Geneva

288. Wellin M (2007) Managing the psychological contract: using the personal deal to increase business performance, 1st edn. Routledge, London
289. Whittaker M et al (2018) AI now report 2018. AI now institute of New York University, New Youk, vol 2018, pp 1–62
290. Wilner A, Babb C (2021) New technologies and deterrence: artificial intelligence and adversarial behaviour. In: Osinga F, Sweijs T (eds) NL ARMS: Netherlands annual review of military studies. TMC Asser Press, The Hague, pp 401–417
291. Wolfson R (2018) Maltese parliament passes laws that set regulatory framework for blockchain, cryptocurrency and DLT. https://www.forbes.com/sites/rachelwolfson/2018/07/05/maltese-parliament-passes-laws-that-set-regulatory-framework-for-blockchain-cryptocurrency-and-dlt/?sh=3180139b49ed. Accessed 02 June 2021
292. Wolfson R (2018) Maltese parliament passes laws that set regulatory framework for blockchain, cryptocurrency and DLT. https://www.forbes.com/sites/rachelwolfson/2018/07/05/maltese-parliament-passes-laws-that-set-regulatory-framework-for-blockchain-cryptocurrency-and-dlt/?sh=59b79a249ed2. Accessed 16 June 2021
293. Wolpow N (2021) What AI founders think about human jobs in and after the pandemic. https://www.forbes.com/sites/ninawolpow/2021/04/27/what-ai-founders-think-about-human-jobs-in-and-after-the-pandemic/?sh=6bb958f51c9a. Accessed 15 June 2021
294. Xiong HY (2015) The human splicing code reveals new insights into the genetic determinants of disease. Science 347(6218):1254806
295. Yager RR, Zadeh LA (1992) An introduction of fuzzy logic applications in intelligent systems, 1st edn. Springer Science+Business Media, New York
296. Yampolskiy RV (2017) Artificial superintelligence: a futuristic approach, 1st edn. CRC Press LLC., Florida
297. Yang X (2019) Accelerated move for AI education in China. ECNU Rev Educ 2(3):347–352
298. Yaqoob I et al (2016) Big data: from beginning to future. Int J Inf Manage 36(6):1231–1247
299. Zakaria M (2012) The role and function of effective communication in the planning phase of international multi-organizational projects, 1st edn. LAP Lambert, s:l.
300. Ziemke T (2004) Embodied AI as science: models of embodied cognition, embodied models of cognition, or both?. In: Lida F, Pfeiffer R, Steels L, Kuniyoshi Y (eds) Embodied artificial intelligence. Lecture notes in computer science, vol 3139. Springer, Berlin
301. Zou J, Schiebinger L (2018) AI can be sexist and racist-it's time to make it fair. https://media.nature.com/original/magazine-assets/d41586-018-05707-8/d41586-018-05707-8.pdf. Accessed 15 June 2021